TECHNICAL REPORT

Network Technologies for Networked Terrorists

Assessing the Value of Information and Communication Technologies to Modern Terrorist Organizations

Bruce W. Don, David R. Frelinger, Scott Gerwehr,
Eric Landree, Brian A. Jackson

Prepared for the Department of Homeland Security

RAND Homeland Security

A RAND INFRASTRUCTURE, SAFETY, AND ENVIRONMENT PROGRAM

The research described in this report was prepared for the United States Department of Homeland Security and conducted under the auspices of the Homeland Security Program within RAND Infrastructure, Safety, and Environment.

Library of Congress Cataloging-in-Publication Data

Network technologies for networked terrorists : assessing the value of information and communications technologies to modern terrorist organizations / Bruce W. Don ... [et al.].
    p. cm.
    Includes bibliographical references.
    ISBN 978-0-8330-4141-8 (pbk.)
    1. Terrorism—Technological innovations. I. Don, Bruce W.

HV6431.N4818 2007
363.3250285—dc22

                                                                2007003787

The RAND Corporation is a nonprofit research organization providing objective analysis and effective solutions that address the challenges facing the public and private sectors around the world. RAND's publications do not necessarily reflect the opinions of its research clients and sponsors.

RAND® is a registered trademark.

Published 2007 by the RAND Corporation
1776 Main Street, P.O. Box 2138, Santa Monica, CA 90407-2138
1200 South Hayes Street, Arlington, VA 22202-5050
4570 Fifth Avenue, Suite 600, Pittsburgh, PA 15213-2665
RAND URL: http://www.rand.org/
To order RAND documents or to obtain additional information, contact
Distribution Services: Telephone: (310) 451-7002;
Fax: (310) 451-6915; Email: order@rand.org

Preface

This report analyzes terrorist groups' use of advanced information and communication technologies in efforts to plan, coordinate, and command their operations. It is one component of a larger study that examines terrorists' use of technology, a critical arena in the war against terrorism. The goal of the investigation reported here is to identify which network technologies might be used to support the activities that terrorists must perform to conduct successful operations, understand terrorists' decisions about when and under what conditions particular technologies will be used and determine the implications of these insights for efforts to combat terrorism.

The information presented in this report should be of interest to homeland security policymakers because it can be used to guide research, development, testing, and evaluation of techniques for collecting counterterrorist intelligence and developing measures to combat terrorism. The results of this analysis may also help inform technology and regulatory policy regarding the development, use, and management of systems that terrorists could use. This work extends the RAND Corporation's ongoing research on terrorism and domestic security issues. This monograph is one in a series of publications examining technological issues in terrorism and efforts to combat it. This series focuses on understanding how terrorist groups make technology choices and respond to the technologies deployed against them. This research was sponsored by the U.S. Department of Homeland Security, Science and Technology Directorate, Office of Comparative Studies.

The RAND Homeland Security Program

This research was conducted under the auspices of the Homeland Security Program within RAND Infrastructure, Safety, and Environment (ISE). The mission of ISE is to improve the development, operation, use, and protection of society's essential physical assets and natural resources and to enhance the related social assets of safety and security of individuals in transit and in their workplaces and communities. Homeland Security Program research supports the Department of Homeland Security and other agencies charged with preventing and mitigating the effects of terrorist activity within U.S. borders. Projects address critical infrastructure protection, emergency management, terrorism risk management, border control, first responders and preparedness, domestic threat assessments, domestic intelligence, and workforce and training.

Questions or comments about this report should be sent to the project leader, Brian A. Jackson (Brian_Jackson@rand.org). Information about the Homeland Security Program is available online (http://www.rand.org/ise/security/). Inquiries about homeland security research projects should be sent to the following address:

Michael Wermuth, Director
Homeland Security Program, ISE
RAND Corporation
1200 South Hayes Street
Arlington, VA 22202-5050
703-413-1100, x5414
Michael_Wermuth@rand.org

Contents

Figures

Tables

Summary

Understanding how terrorists conduct successful operations is critical to countering them. It has become apparent that terrorist organizations are using a wide range of technologies as they plan and stage attacks. Most examinations of the technology used to enable terrorist operations focus on their weapons—the instruments directly responsible for death and destruction in their attacks—and how new technologies might increase the resulting damages, injuries, and fatalities. However, successful terrorist operations involve more than simply employing weapons to produce their physical effects. Information gathering, assessment and planning, coordination, logistics, and command capabilities all play a role in delivering the terrorist's weapon to its intended target with deadly effect, and the very existence of a terrorist organization is based on recruiting and information campaigns. As a result, understanding the role that such technologies play and the net effect of their use requires an understanding not only of the technology, but also of the purpose and manner in which the technology is used and of the operational actions and responses of the security forces and the terrorists. To gain such an understanding, the study has taken a broad scope in assessing the issue.

Study Scope and Purpose

This analysis focuses on the potential application of information and communication technologies that may be used across the full range of activities that make up terrorist operations and whether these applications can lead to new and different approaches to terrorist operations. Its purpose is to identify which of these network technologies terrorist organizations are likely to use in conducting their operations and to suggest what security forces might do to counter, mitigate, or exploit terrorists' use of such technologies.

To highlight the merger of software and computer technologies with communication and display technologies that digitalization has made possible and to encourage thinking beyond military technologies, this report uses the term *network technologies* to describe what are referred to as command, control, communication, computer, intelligence, surveillance, and reconnaissance (C4ISR) technologies in military parlance, as well as the consumer-oriented technologies that can often provide the functionality needed for terrorist operations. These network technologies can include connectivity technologies (e.g., wireless routers), mobile computing (e.g., laptop computers), personal electronic devices (e.g., personal digital assistants and cell phones), IT services and Internet access, and video recording, among others.

Approach to the Analysis

The RAND research team used five research questions to guide the analysis of the terrorist use of network technologies and to identify effective ways for security forces to counter their use.

1. What could terrorists do with network technologies?
2. Which network technologies are most attractive to terrorists?
3. How would specific network technologies fit within terrorist groups' broader approaches to acquiring and using technologies?
4. What should security forces do to counter this?
5. What conclusions and recommendations can be drawn from this analysis?

First, the team developed a terrorist activity chain shown in Figure S.1. It is a logic model that describes the activities that make up most terrorist operations and explains how these activities relate to one another.

Next, the team examined terrorist use of network technologies for the elements of the terrorist activity chain to discover which of the activities could benefit from terrorist use of network technologies and which network technologies might promise the most substantial benefits. To do this, the study team based its investigation on the following questions:

Figure S.1
The Terrorist Activity Chain

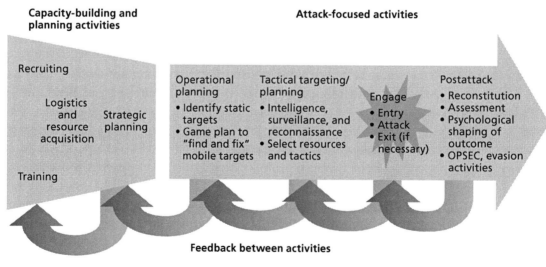

NOTE: OPSEC = operational security.

RAND *TR454-S.1*

- How have terrorists used network technologies to support terrorist operations *in the past?*
- How are terrorists now using network technology to support their *current operations?*
- What uses of network technologies may terrorists be *expected to make in the future*, and might such use lead to revolutionary changes in future operations?

The next step was to identify which network technologies were most attractive to terrorists. The team analyzed the types of network technologies that would be most useful for a given terrorist activity, whether they would be practical to acquire, and whether any technologies might offer revolutionary changes. We base our assessment on the expectation that terrorists will adopt a technology if it can confer one of two types of benefits with reasonable risks:

1. those that improve the organization's ability to carry out activities relevant to its strategic objectives, such as recruiting and training, or
2. those that improve the outcome of their attack operations.

The team then developed a structured way of thinking about how terrorists acquire technologies and the role that specific network technologies play within groups' technology strategies. These technology strategies are as follows:

1. *Invest in specialized technology, in pursuit of a significant effect on attack outcomes or perhaps operational efficiency.* Typically, such technologies require some parts of the organization to specialize for effective acquisition and employment.
2. *Either rely on versatile technologies that can be used many ways or pursue a wide variety of individual technologies, with the expectation of a moderate effect on operational efficiency and, perhaps, some positive benefits for attack outcomes.* Groups frequently acquire technologies relevant to both these strategies externally from legal or illegal market sources.
3. *Use technology opportunistically, with the expectation that technology will only contribute to attack outcomes and operational efficiency in minor ways.* Such a strategy may also result in little organizationwide vulnerability to technology failures, countermeasures, or exploitation.

These strategies summarize the approaches that have been successful for terrorist organizations in light of the basic characteristics of *both* the technology and the manner in which it could be used. They crudely incorporate a broad set of factors that are fundamentally related to one another: the nature of the technology, the operational environment in which it would be useful, the general effect of its use, and the acquisition approach it requires. As a result, they provide a simple model that can serve as a framework for evaluating the effectiveness of alternative ways for security forces to respond to these general approaches to technology by a terrorist organization.

Finally, the team evaluated how to best counter terrorists' use of network technologies. This required the research team to assess and compare the benefits and risks of different countermeasure options. To do this, we developed a framework that considers three basic factors:

1. the role that a specific network technology plays within a terrorist group's overall technology strategy
2. the balance of benefits and risks of technology use from both the terrorists' and security forces' perspective
3. options for security forces to counter terrorists' use of network technologies.

This framework allowed the team to compare the payoff for each combination of network technology used by terrorists and countermeasure available to security forces.

As any analysis, this approach has its limitations. Because terrorists will not necessarily use technology or conduct operations in the ways that they have in the past, the conclusions of this analysis are limited most importantly by how insightful the research team has been in two areas: envisioning how clever terrorists can be in their future use of network technology and understanding the limitations that realistically constrain future terrorist operations. Unforeseen new uses are certainly possible, given the rapid pace of technology development, and future operations involving terrorists may be very different from current operations. However, the team believes that the approach we have used for this analysis is uncomplicated and flexible enough to be used on a continuing basis to examine startlingly new or evolving situations. This need for update and review is the basis for our recommendation suggesting that DHS put in place a system to do this on an ongoing basis.

Conclusions

Future network technologies are most likely to result in real but modest improvements in overall terrorist group efficiency but not dramatic improvements in their operational outcomes. This results largely from the circumstances under which terrorist groups must operate, particularly in the homeland security arena, and the carefully planned and scripted style of their attacks. These groups must operate through inherently fragile, clandestine terrorist cells that have resource limitations, a need for secrecy for survival, and a need for surprise and scripted attacks for operational effectiveness. All of these considerations result in an operational style that favors uncomplicated operations with concrete effects and minimal core needs for the capabilities that network technologies provide.

Terrorists will most likely acquire network technologies for the versatility and variety that they offer and will use them to enhance the efficiency and effectiveness of their supporting activities. The effect of these kinds of technologies will be to make their activities more efficient or effective. That is, they will be able to carry them out with fewer people or better results. Thus, they might be able to get by with fewer people devoted to recruiting new members because one person might be able to recruit more new members.

Attempting to preclude terrorists from getting the types of technology they want will not be practical, and developing direct counters to them will unlikely yield a high payoff. Network technologies that feature versatility and variety are largely driven by the worldwide consumer and commercial markets. It is not practical to keep these kinds of technologies out of the hands of terrorists. Such technologies can simply be bought off the shelf. Even if it were possible to deny terrorists these technologies, the benefits of doing so would probably not justify the costs of the effort required to block their acquisition.

Exploitation seems the more promising option. The best use of resources for those attempting to counter terrorist operations would seem to be developing ways to exploit the network technologies that terrorists will continue to use. As is the case with most people who use cell phones and computers, most terrorists do not have detailed knowledge of how those devices work. Therefore, it may be possible for sophisticated security forces to alter them in ways that enable security services to identify the users or their locations or to monitor their transmissions. This approach also targets a key vulnerability: an absolute need of terrorist organizations to remain hidden.

Even though there do not appear to be any network technologies that offer revolutionary capabilities in the immediate future, security services need to monitor the development of technologies in the event that such a capability emerges. One area that might require careful monitoring would be network technologies that enable terrorist organizations to assume the identity of government personnel (perhaps electronically) or take over media outlets. Even though it is unlikely that they could do this for a sustained period, even a short takeover could be terribly disruptive, particularly in densely populated urban areas.

Recommendations

In light of the above conclusions, the research team recommends the following actions.

Design a system to address terrorist use of network technologies. Security organizations need a process that determines whether new network technology has been or is likely to be introduced into terrorist operations, identify its effect, select a response, gather needed resources, and implement an appropriate counter to the technology's use, and to do all of these in a timely manner.

Acquire and sustain people with the core competencies needed to make the system work. Homeland security forces and other organizations involved in combating terrorism need the following core competencies to address the use of network technologies by terrorist organizations:

- an understanding of the technologies themselves, particularly the technical challenges of exploitation and the operational limitations imposed by terrorist and security force operations
- an ability to track terrorist adoption, use, or avoidance of particular technologies
- a capability to determine which responses, or which mix of responses, is most appropriate in light of security force goals, and

- the capacity to develop plans and execute operations to actuate the selected responses as part of the larger strategy to counter terrorist organizations.

Take the initial steps needed to implement such a system promptly. Initial actions that can quickly provide a good basis for a system that can counter terrorist organizations' network technology use include the following DHS activities:

- Continue and accelerate the recruitment, retention, and professional education of technically skilled personnel who understand network technologies.
- Define the requirements for intelligence collection that focuses on terrorist use of network technologies and communicate them to the intelligence community.
- Create an effort within the homeland security research program to examine terrorist use of network technologies.
- Develop the capability to determine whether to exploit the use of the network technology; develop and employ operational countermeasures to the network technology; disrupt the process by which terrorist groups acquire new network technologies; or determine that other counterterrorism efforts are more effective than a response.
- Develop a capability to respond quickly with technical and engineering solutions to counter or exploit emerging network technology being used by terrorists.

These actions should provide a basic capability within DHS that can contribute to the homeland security mission in the short term and that can be shaped to provide the most efficient and effective ways to address this threat over the longer term.

Abbreviations

BR-PCC	Brigate Rosse per la Costituzione del Partito Comunista Combattente
C4ISR	command, control, communication, computer, intelligence, surveillance, and reconnaissance
CDMA	code division multiple access
DARWARS	U.S. Defense Advanced Research Projects Agency's universal, persistent, on-demand, training wars
ETA	Euskadi Ta Askatasuna, or Basque Homeland and Liberty
FARC	Fuerzas Armadas Revolucionarias de Colombia, or Revolutionary Armed Forces of Colombia
FLN	Front de Libération Nationale
GIS	geographic information system
GSM	global system for mobile communication
IED	improvised explosive device
IRC	Internet relay chat
LTTE	Liberation Tigers of Tamil Eelam
MIT	Massachusetts Institute of Technology
MMOG	massively multiplayer online game
MRTA	Movimiento Revolucionario Túpac Amaru
OPSEC	operational security
PGP	pretty good privacy
PIRA	Provisional Irish Republican Army
RDD	radiological dispersal device

RFID	radio frequency identification
RIRA	Real Irish Republican Army
SANS	SysAdmin, Audit, Network, Security
SMS	short message service
VOIP	voice over internet protocol
VPN	virtual private network
WiFi	wireless fidelity (IEEE 802.11x wireless networking)

CHAPTER ONE

Introduction

Understanding what contributes to the success of terrorist operations is critical to countering their attacks. Terrorist organizations are using a wide range of technologies as force multipliers as they plan and stage attacks. These technologies range from the relatively simple adaptation of garage-door openers to detonate explosives as targeted vehicles pass by to the sophisticated development of videos or Web sites to trumpet terrorist successes or to recruit new members. Technology, of course, does not stand still. Global consumer demand for new capabilities or products has fueled an explosion of new or enhanced technologies, many of which terrorists could use to make their operations more efficient or effective. However, technology can be a double-edged sword: As it boosts effectiveness or efficiency, it might also introduce new vulnerabilities. Thus, the terrorist's choice of whether to adopt a new technology is not necessarily straightforward, which makes it difficult for security services to know to which future technologies they should respond and what would constitute an appropriate response when one is necessary.

The Scope and Purpose of the Analysis

The analysis in this report focuses on the potential use of information-based technologies by terrorist organizations in their activities. The purpose is to identify which of these technologies terrorist organizations may find attractive for carrying out their operations and to suggest what security forces might do to counter, mitigate, or exploit the use of such technologies.

Terrorists use many different types of technology. In this report, we focus on what we call *network technologies*. These information-based technologies include what might be described as the canonical military command, control, communication, computer, intelligence, surveillance, and reconnaissance (C4ISR) technologies[1] as well as the consumer-oriented technologies

---

[1] These include technologies used for command, control, communication, computation, intelligence collection and analysis, surveillance, and reconnaissance. The study team has avoided describing the technologies of interest simply by reference to their military analog (C4ISR) because of its view that this can limit the analysis by casting terrorist organizations as military units without uniforms. Although terrorists rely on the same types of information that C4ISR systems are designed to provide, the information that terrorists need and their method of acquiring it are markedly different from the organized military's information and methods. For fundamental reasons (our open society, the difference in military versus civilian targets, and the size and operational profile of security forces), information about security forces and terrorists' targets is often easy to collect because it is readily available and often apparent. The necessary information can be collected by persons

that can provide the functionality needed for terrorist operations. They help store, communicate, manipulate, and display information. Network technologies can include the following:

- connectivity technologies (wireless communication modes)
- mobile computing
- personal electronic devices (e.g., PDAs, cell phones)
- software and applications
- IT services and access to the Internet
- video and other recording devices.

Although these technologies can aid terrorist organizations by enabling military functions like command and control (see, for example, Whine, 1999), they can also provide capabilities that increase terrorists' effectiveness in other necessary activities such as raising money or persuading people to join their causes.

Research Approach

The approach the research team used is based on a series of five questions:

1. What could terrorists do with network technology?
2. Which network technologies are most attractive to terrorists?
3. How would specific network technologies fit within terrorist groups' broader approaches to acquiring and using technologies?
4. What should security forces do to counter this?
5. What conclusions and recommendations can be drawn from this analysis?

The following sections explain the approach in more detail.

What Could Terrorists Do with Network Technology?

As a first step in understanding what other uses terrorists might have for network technologies, we needed to develop a structured way to think about what terrorists do. Describing terrorist activities may, at first, seem obvious, as terrorist operations involve attacks against people who have little ability to defend themselves. But the attack itself is only part of what a terrorist organization must do to succeed; in addition, many activities before and after an attack can spell success or failure, particularly over the course of an extended terrorist conflict.

Although it is tempting to use a military operational model to define terrorist activities, applying such models is difficult because, in terrorist organizations, a small group typically carries out the functions of an entire military establishment. Moreover, many of the approaches

---

with little experience or training through the use of consumer electronics such as video recorders or cameras. In contrast, military forces seeking to obtain analogous information must often rely on complex systems because their adversaries go to great lengths to hide or protect critical information.

used for basic terrorist activities are much different when conducted in the terrorists' clandestine environment from those carried out in the domestic environment of a nation-state.

To parse what a terrorist organization must do to succeed and how terrorists might use network technology to help with those activities, the research team developed the terrorist activity chain as shown in Figure 1.1. It is a logic model that describes the activities that make up most terrorist operations and how these activities relate to one another.

To execute operations and sustain itself over the long term, the terrorist organization must succeed at each of the broad tasks listed in the figure; these tasks include both capacity-building and attack-related activities. We describe each below.

- *Recruiting:* This is the process of attracting motivated individuals with the right skills and capabilities to the terrorist's cause.
- *Training:* This provides organization members with a way to learn new skills and refine them over time. Such learning requires more experienced members to transfer knowledge to newer members and encompasses both individual skills and unit abilities.
- *Acquiring financing and physical resources:* An organization amasses whatever resources are needed to sustain it and its operational and support activities. Depending on the group's plans and strategy, resource requirements may vary from modest to more extensive and include physical assets such as weapons and financial assets.

Figure 1.1
The Terrorist Activity Chain

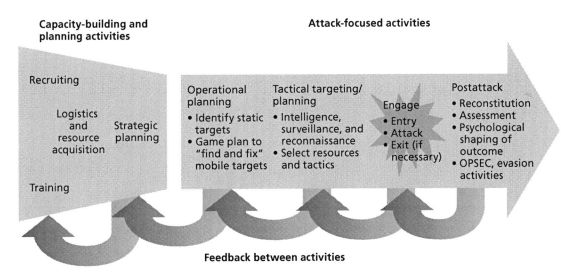

NOTE: OPSEC = operational security. This particular model of terrorist activities was developed by the RAND project team and is similar to other organizational activity models found in the literature (see, for example, U.S. Army Training and Doctrine Command, 2005). The activity chain was used to provide a framework for analysis of the technologies in this study and to provide a common reference point for other technology-focused projects that were being carried out as part of this research effort. The results of those projects appear in separate publications.
RAND TR454-1.1

- *Developing a strategy:* Terrorist actions are intended to accomplish political or social goals. To guide its actions, a group must develop a strategy to link these goals to specific actions. In some cases, such as a terrorist group affiliated with a larger movement such as al Qaeda, the group's strategy may be provided exogenously—that is, by the parent group.
- *Identifying targets:* Modern society presents terrorist groups with a wide variety of potential targets, ranging from specific individuals, members of the public, critical infrastructures and installations, and symbolic sites. Because the attack method and effects that are best for one target may differ distinctly from the approach that is most effective for another, most terrorist groups invest time and resources in identifying and choosing targets that their leadership believes best suit their purposes and capabilities.
- *Planning operations:* To carry out a terrorist operation, a group must gather the intelligence needed to attack a selected target. Human and technical resources must be allocated to the attack, roles, and timing defined, all appropriately matched with the security and operational constraints that must be overcome for the mission to succeed.
- *Conducting attack operations:* At the point of attack, the terrorist must successfully approach the target, engage it, and, if the operation is not designed to result in the death by suicide of the operatives involved, escape.
- *Shaping public reaction and preparing to conduct subsequent attacks:* After an operation, any terrorists who remain must escape and continue the organization's activities. Because much of a terrorist attack's effect is determined by public reaction, the organization may undertake postattack actions such as claims of responsibility and other messages to the public or the authorities to ensure that the message the group intends to convey reaches a wide audience, thereby increasing its benefit to the group. To prepare for future attacks, the group must reconstitute its capabilities and begin anew the sequence of activities shown in our activity chain model.

By examining terrorist use of network technologies for these elements of the terrorist activity chain and by comparing this to expected network technology capabilities for the future, we can discover which activities would benefit from network technologies and which network technologies might promise the most substantial benefits. To do this, the study team next looked for trends and important discontinuities through the following questions:

- How have terrorists used network technologies to support terrorist operations *in the past?*
- How are terrorists now using network technology to support their *current operations?*
- What uses of network technologies may terrorists be *expected to make in the future,* and might such use lead to revolutionary changes in future operations?

Specifying the activities, considerations, and objectives for each of the tasks necessary for terrorist organizations in such an activity chain provides the basis for systematically assessing for what functions terrorists might use network technologies and to what network technologies terrorists might be most attracted.

Which Network Technologies Are Most Attractive to Terrorists?

To answer this question, we needed a basis for systematically exploring how terrorists evaluate a new technology in light of its basic characteristics and potential uses. We used a model of terrorist decisionmaking that posits that the group adapts to the operational situation it faces to survive and be successful in its mission. As a result, we base our assessment on the expectation that terrorist groups adopt a technology if it can confer one of two types of benefits with reasonable risks:

1. those that improve the organization's ability to carry out activities relevant to its strategic objectives, such as recruiting and training, or
2. those that improve the outcome of its attack operations.

How Would Specific Network Technologies Fit Within Terrorist Groups' Broader Approaches to Acquiring and Using Technologies?

To answer this question, we rely on research by Jackson (2001) and Jackson, Baker, et al. (2005a, 2005b) that analyzes the basic actions that a group must carry out to adopt a new technology and assesses organizational learning in terrorist organizations. We use the concept of technology strategies to define a simple framework to summarize the approaches that terrorist groups take in acquiring and using network technologies. The resulting framework summarizes four broad approaches that terrorist groups take with respect to new technologies:

*   *Specialize in specific technologies, enabling the group to customize and shape them to the needs of its activities and operations.* Typically, implementing such an approach requires some parts of the organization to specialize for such technology to be acquired and used.
*   *Adopt many technologies, providing the group with a wide variety of options to apply as needed.* Although variety-based strategies do not necessarily require groups to build up specialization or deep knowledge of particular technologies, groups must invest time and resources in maintaining their ability to use many different technologies well. Variety-based strategies are made much easier when technologies are readily available on the commercial market.
*   *Focus on individual technologies, but choose ones that are versatile and can be used many different ways.* The more ways in which an individual technology can be used, the higher its potential value to an individual terrorist group. The ubiquity of communication across the terrorist activity chain—and the availability of these technologies on the commercial market—demonstrates that many network technologies could constitute very versatile technologies within these groups' operations.
*   *Rely on technology opportunistically, without a concerted organizational focus on adopting and deploying novel technologies.* Just because technologies appear potentially attractive to terrorists, there is no certainty that they will adopt them. Although passing up opportunities to use new technologies will deny organizations their benefits, such a strategy may also result in little organizationwide vulnerability to technology failures, countermeasures, or exploitation.

What Should Security Forces Do to Counter This?

Given what we learn from the analysis of future network technologies and how terrorists might acquire and use them, the next step is to assess what options are available to security forces.

This question requires the research team to assess and compare the benefits and risks for different countermeasure options. To do this, we developed a method for determining the value of different countermeasures in light of the technology strategies that terrorists use. The framework used for this has three basic components:

- the role that a specific network technology plays within a terrorist group's overall technology strategy
- the balance of benefits and risks of technology use from both the terrorists' and security forces' perspective
- options for security forces to counter terrorists' use of network technologies.

When used together, these components can define a framework that allows us to compare the payoff for each combination of network technology used by terrorists and countermeasure available to security forces.

What Conclusions and Recommendations Can Be Drawn from This Analysis?

This analysis leads to conclusions and recommendations in three broad areas:

1. What changes in terrorist operations and their outcomes are network technologies likely to enable in the future? Are there any truly revolutionary capabilities that may develop?
2. What are the broad characteristics of effective ways to counter the advantages that terrorists may derive from such technologies? How should we deal with unexpected advantages that terrorists may develop?
3. What actions should DHS and other security forces take in light of the insights that this report provides? Are there any hedging activities to guard against revolutionary surprises?

The conclusions of the analysis are limited by how insightful the research team has been in two areas: envisioning the ways in which terrorists can use network technology and understanding the limitations that constrain their future terrorist operations. These are, of course, not technical limitations; they relate directly to the issue that the National Commission on Terrorist Attacks upon the United States (the 9/11 Commission) referred to when it cited a failure of imagination as one of the prime shortcomings in U.S. ability to prevent the attacks from happening (National Commission on Terrorist Attacks upon the United States, 2004, p. 336). Unforeseen new uses of network technology by terrorists are certainly possible with the pace of technology development today. Similarly, future operations involving terrorists may be very different from current operations with very different operational constraints and perhaps very different objectives.

As a consequence, it would be prudent to hedge against failures of imagination. The team believes that the approach it has used for this analysis is uncomplicated and flexible enough to be used to reexamine key aspects of this issue on an ongoing basis. This need for update and review (and the consequent changes to programs and strategies) is the basis for the third category of recommendations outlined above, which includes the suggestion that DHS put in place a system to examine this issue as part of its regular activities.

How This Report Is Organized

This report has four chapters, including this introduction. The bulk of the analysis appears in Chapter Two, in which we describe the network technologies that terrorists might want to acquire, for what they are likely to use them, and which network technologies that appear to be most attractive from their perspective. Chapter Three provides an analysis of the possible responses that security forces could take to counter their acquisition and use. Chapter Four provides the study's conclusions and recommendations.

CHAPTER TWO

What Could Terrorists Do with Network Technology?

To assess how specific network technologies would affect terrorist groups' operations, we based our analysis on the activities that terrorist groups must accomplish to successfully execute their operations and sustain their effort over time. These activities range from capacity building to postattack operations. Using the terrorist activity chain developed in the previous chapter, the research team selected nine basic terrorist functions that depend significantly on network technologies; the expanded version of the terrorist activity chain in Figure 2.1 depicts them:

1. recruiting
2. acquiring resources
3. training
4. creating false identities, forgery, and other deception
5. reconnaissance and surveillance
6. planning and targeting
7. communication
8. attack operations
9. propaganda and persuasion.

In the following sections, we define each of these nine basic terrorist group functions and assess the potential effect of network technologies on terrorist activities. This analysis is informed by how terrorists have carried out these activities in the past, but it also takes into account the current state of the art, both technical and operational, as well as likely future technical capabilities. Because enhanced technological capabilities can bring about entirely new ways of doing things, the study team also examined the potential for network technologies to bring about revolutionary changes in terrorist capabilities and operations.

Recruiting

We define *recruiting* as the process and tools that an organization uses to attract and indoctrinate new members. New members are essential for terrorist groups, because members are killed or arrested, defect, or simply lose interest in the cause.

Figure 2.1
The Basic Functions of the Terrorist Activity Chain

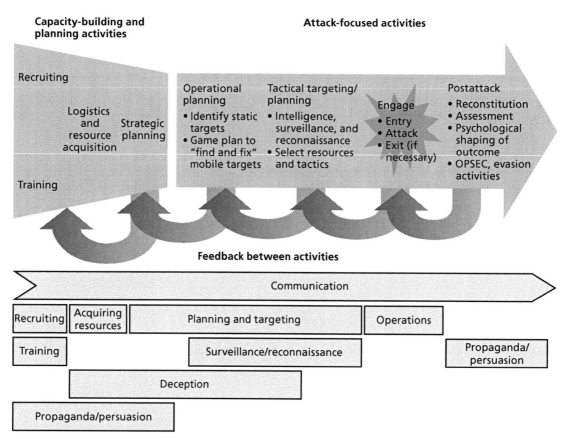

NOTE: As with any model, our activity chain is a simplification of a more complex reality. As such, although it serves its purpose for this report, it is limited in how literally it can be used for other purposes. For example, although we depict training, which can represent initial basic training for recruits, at the far left of the figure, the feedback loops are intended to imply that group members can enter into training from any point along the chain (and, therefore, as more experienced combatants). Additionally, training activities can be accomplished at several points along the chain (e.g., practicing for an attack operation is a form of training). Although we have located the training bar to the left of the figure to provide some sense that it usually precedes the other activities, training could be considered to run throughout the chain, as we have depicted communication.

RAND *TR454-2.1*

Initial recruiting efforts seek to identify and gain access to populations suitable for recruiting.[1] For example, a group with a religious ideology might take steps to ensure that it has access to congregations and religious schools and subsequently seek to influence the sermons or curricula within those institutions. Such access has been important in the recruitment into groups in Pakistan including groups related to al Qaeda and to extremist groups focused on local agendas.[2] Another example of access is the role that Islamic religious schools, or madras-

---

[1] A suitable population is one that is "available." That is, the population is experiencing that combination of social, cultural and other environmental variables that makes it receptive to recruitment attempts.

[2] See, for example, the discussion of al Qaeda recruitment in Pakistan in Fair (2004).

sas, have sometimes played in the identification and indoctrination of potential members of Islamist terrorist organizations in many nations.[3]

The next step in the recruiting process is the first contact between the organization and a nonmember, whether direct (e.g., a face-to-face meeting) or mediated (e.g., a Web site posting or a meeting with a friendly member of the clergy). What normally follows are incremental steps in an indoctrination process. Recruiting and indoctrination activities continue through a phase called *developed contact*, in which the candidate's attitudes are confirmed or reshaped to fit the group's doctrine.[4] The recruiting process ends once indoctrination is complete. At this point, the individual self-identifies as a member and becomes involved in the organization's activities.

The recruiting process may vary widely from organization to organization and even within an organization. For example, individuals may be recruited on the basis of demographics (e.g., gender, nationality, age), skill sets, family, and other social connections, or purely by opportunity. Aum Shinrikyo, the Japanese terrorist group responsible for the 1995 sarin gas attack on the Tokyo subway, provides an example of the latter approach to recruiting (Parachini, 2004). First-person narratives of members of the Provisional Irish Republican Army (PIRA) reveal, for instance, that individuals were recruited and indoctrinated into specific organizational functions (Collins and McGovern, 1998; O'Callaghan, 1999). Once in the recruiting pipeline, individuals may be assigned to particular roles[5] or gravitate naturally to them, or chance may dictate their ultimate position.[6]

Recruiting normally involves employing a wide variety of communication methods— videos, pamphlets, Web sites, sermons, friendly news media, personal friends, and other influential people—in a number of locations: private homes, schools, religious sites, paramilitary camps, prisons, and so on. These aspects can be used to define two basic dimensions of recruiting:

- *Public versus private channel.* Is the interaction taking place in or out of the public eye? The prevailing laws of the region, rules of the local institutions, and attitudes toward the group all will greatly affect where recruitment efforts fall on this spectrum.
- *Proximate versus mediated contact.* Is the source of the recruitment effort physically close to the target audience? Cultural, technology, and economic circumstance are some of the variables that influence how the recruiting message can be passed to the intended target audience.

Figure 2.2 (derived from Goffmann, 1963) illustrates these two cardinal dimensions of recruiting interaction: public versus private and proximate versus mediated. The rapid proliferation of network technology greatly increases the opportunity for interactions in mediated recruitment and for effective interactions in proximate recruiting efforts.

---

[3] See, for example, discussion of recruitment practices by Jemaah Islamiyah in Baker (2005).

[4] The model invoked was first presented in Zimbardo and Hartley (1985).

[5] See discussion of the Real IRA's specific recruitment of bombmakers in Cragin and Daly (2004, p. 27).

[6] See, for example, Jackson (2006b), describing PIRA's winnowing process for specialists within the group.

Figure 2.2
Cardinal Dimensions of Recruiting

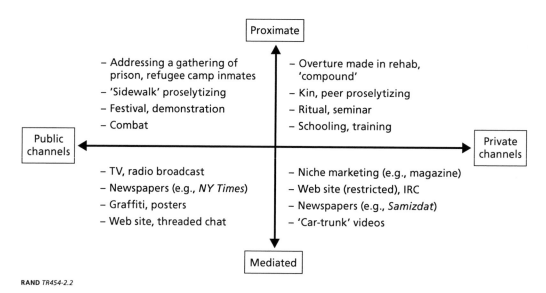

RAND *TR454-2.2*

Historically, recruiting for terrorist organizations has been a clandestine process. The need for security and secrecy heretofore has necessitated a low profile and often required that it be conducted face to face.[7] Terrorist organizations' causes have traditionally been—and still are in many cases—local and parochial.[8] Parochial causes generally have smaller pools of people from which to draw new recruits; there are simply fewer individuals who care about narrow causes than those who care about broader ones. Face-to-face recruiting limits the number of individuals who can be contacted. Moreover, small-scale recruiting coupled with the need for secrecy generally has meant a longer recruitment process, as the process must take place unobserved by security (often at a single site or in a few locations). Finally, recruitment into terrorist groups has frequently involved a lengthy proving period. In such circumstances, the technology available and the nature of the recruiting activity both worked to keep the cause local and the pool of potential recruits limited.

Current State-of-the-Art Recruiting

Today, forms of recruiting enabled by network technology greatly expand the scope, effectiveness, and efficiency of previous recruitment activities. First, recruiting can be done remotely. With recruiting materials on the Internet available from almost anywhere, face-to-face contact is not a necessity. This can facilitate recruiting by making a broad audience aware of a group's existence and cause. Second, remote recruiting is efficient because a single recruiter can develop many candidates at the same time. Terrorist recruiters may now simultaneously work with audiences in many parts of a single country or in many far-flung countries, expanding the pool of potential recruits. For example, Hizballah has used a number of violent video games

---

[7]   See, for example, the Anti-Defamation League study on extremist recruitment in prisons (B'nai B'rith, 2002).

[8]   The Occupied Territories and East Timor are examples.

with names such as *Special Force* and *Under Ash* as part of its effort to get its pro-Palestine and anti-Israel messages across and to attract new recruits in Lebanon as well as abroad (Harnden, 2004; Lewis, 2005).

Current network technologies, such as Internet access, networks, and video games can increase a group's ability to spread its message broadly, often with the message tailored to particular target audiences. They can also allow recruiters to operate from a safe haven, out of reach of security forces in the targeted countries.

The Future of Recruiting

Increasingly, data on individuals are being collected and warehoused in electronic form. These data can often provide very detailed information on such matters as purchasing habits and personal tastes (see, for example, Thibodeau, 2001). This is a global phenomenon, not a practice confined to the United States. Such data warehouses may be exploited for the purposes of recruiting by terrorist organizations. Such recruiting tactics could increase the efficiency and effectiveness of recruiting activities in the future by allowing recruiters to target individuals likely to be sympathetic to a terrorist organization's message just as marketing organizations attempt to do. It can also be sufficiently specific as to location to aid terrorist recruiters trying to develop recruits in a particular place or region. The personal profiles developed by search engines have a similar potential to identify individuals who may have an interest in a group's message. Both of these techniques require that the terrorist organization's recruiters have access to such personal information; however, whether personal information is acquired by hacking, pretexting, or merely buying data, acquisition has not proven to be a major impediment for others seeking to use such information and is not likely to constrain its use for terrorist recruiting.

Limiting a terrorist's ability to recruit new members is already difficult. However, some technological advances might make countering terrorist organization recruiting harder still. Recruiting could be made more effective and efficient by the transfer of all or most of the indoctrination process into a virtual setting (e.g., online, videos). Although much recruiting may already be done virtually, indoctrination is more problematic, since many of the techniques used in indoctrination typically require immersion and proximity.[9] Indoctrination is facilitated by isolation of individuals from conflicting ideas and information in a way that makes more complete immersion into group ideology possible. Carrying out indoctrination processes through virtual channels would require that individuals be willing to isolate themselves, even in the absence of direct control over their actions by group leaders. Shifts in both technologies and how people relate to those technologies could make it easier for such "indoctrination at a distance" to occur, but would require that the technology create a compelling experience.

The latest generations of computer-based, massively multiplayer online games (MMOGs), in which many individuals interact in a common virtual world, constitute a step toward the

---

[9] For a discussion of recruitment models, see Daly and Gerwehr (2006, pp. 76–80) and Cragin and Gerwehr (2005, pp. 19–20). Also, Ramakrishna (2004) describes the intense and hands-on indoctrination process within Jemaah Islamiyah.

conditions in which such indoctrination might take place.[10] Several factors suggest the utility of such virtual worlds to terrorist groups seeking mechanisms to support recruiting and indoctrination. First, the games are engaging; individuals willingly invest large amounts of time in their activities in these virtual worlds. Second, participants often form tight relationships with each other within the framework of computer games, even if they have never had a face-to-face contact. Third, players find the games sufficiently compelling that the boundaries between the virtual world and the real world blur; players sometimes spend real-world money to purchase properties in virtual worlds, conflicts between players that are linked to in-game events arise, and so on (Patrizio, 2002; Loftus, 2005; Yee, 2006a, 2006b).

At first glance, it might appear that MMOGs and online gaming might be a boon to terrorists around the world, in part because of the high degree of communication and interactivity that these games enable. However, as intriguing as the games are and the possibility is that they could be used in ways to help in some serious applications such as reinforcing principles learned in conventional training situations, they represent a fairly modest enhancement to the terrorist repertoire of communication techniques. The communication enabled inside the game does not differ not significantly from other Internet-enabled communication, except that it might go unmonitored by security services focused on current Internet communication media such as Internet relay chat (IRC), chat rooms, or voice over internet protocol (VOIP). In many ways, the types of communication that such games enable are variants of these media milieus rolled into a single package with reinforcing graphics. Like the other forms of electronic communication, many issues are associated with engaging in clandestine communication that would give a thinking adversary pause before using any system not under its control for sensitive communication. For instance, servers represent a meeting point in many of these games and are a major point of vulnerability for the would-be communicator, as are the open nature of many of the games that enable players to join the game. Communication on these machines could certainly appear to be private, but environments such as multiplayer games on the Internet offer vulnerabilities that security forces could exploit.

The vulnerabilities of game communication can result from server-side exploitation, interception of the packets moving through the network, end-point vulnerability (compromise of the computer connecting to the server), or user compromise. The adage "you can't tell if it is a dog on the Internet" is both humorous and true. Ensuring privacy, or security from the terrorist recruiter's point of view, usually requires the exchange of additional information through an entirely separate communication method to establish identity and a degree of trust. In practice, this sort of operation is possible, but it is difficult to do well, or with reasonable risks, for groups under pressure.[11]

A more interesting element of MMOGs, however, is that they might be a means by which groups may begin associations that they take offline, and thereby become a means of helping

---

[10] For a brief overview of culture and social networks in massively multiplayer games, see Jakobsson and Taylor (2003).

[11] As is the case for all secret communication techniques, the use of the MMOG has its strengths and weaknesses. In particular, this technique requires some additional way for parties removed from each other to authenticate themselves to one another. Consequently, the use of the MMOG for secret communication is a clever technique but not a dramatically new capability for a group that is able to manage its secret communication properly.

in recruiting processes. In this role, the secret communication elements are not important, but rather the affiliation itself is, as a stepping stone for other activities. This offers terrorist recruiters a way of meeting people in a setting that is not overtly associated with their groups' message and could act as a cut-out that might go unobserved by security forces. From the security forces' perspective, this means that such game sites might be good targets to monitor. Much as they do in the real world, virtual-world activities pose opportunities and challenges for both the terrorists and security forces.

From the security force side of the problem, MMOGs may appear daunting, not only because they generate yet another large stream of data with which to deal, but also because they could represent a potentially embarrassing element of a terrorist plot that, in hindsight, might have been easily discovered. However, if security force efforts are guided by additional intelligence information, these can be exploited to provide a potentially useful window into a terrorist group and its activities.

Acquiring Resources

Acquiring resources is the act of obtaining physical assets, information, and money needed to conduct terrorist operations.

Traditionally, international terrorist groups have relied on criminal activities or international donations to acquire resources.[12] Although technology has played some role, it has been limited; terrorist organizations have relied more heavily on complex organization, functional differentiation, and specialized skills for successful financing.

Current State-of-the-Art Resource Acquisition

The common use of technology in everyday affairs has changed the acquisition landscape that terrorist groups occupy. This landscape now includes the use of technology to enhance group criminal and psychological activities (Emerson, 2002), and it makes financial tools such as cyberpayments and Internet banking as well as money laundering and other financial crimes increasingly available for terrorist use (Wilson and Molander, 1998). Examples of such exploitation include use of a computer and coding device to alter and create credit cards (*United States v. Mokhtar Haouari*, S4 00 Cr. 15 [JFK], S.D.N.Y., July 3, 2001, p. 563) or using electronic transfers of funds to lower exposure and eliminate the risk of physical contact. Network technology also facilitates the use of technology-enabled informal banks such as hawalas that are widely used to transfer funds between individuals outside formal financial systems. Because transactions made through such systems do not require either face-to-face interaction or travel and they have often been lightly monitored and audited, they permit terrorist organizations to exercise a global reach at relatively low cost with relatively low risk.[13]

---

[12] Adams (1986); for a detailed discussion of a single terrorist group, the PIRA, see Horgan and Taylor (1999, 2003).

[13] Hawalas are unregulated international money transfer networks—*hawala* means "in trust" in Hindi. Immigrants in developed countries commonly use them to transfer cash locally or abroad to people who do not have access to the formal

Network technology has also enabled greater advantage in propaganda and other information operations that are part of terrorists' resource acquisition efforts; the worldwide audiences for the 24-hour television news services as well as the Internet have provided a ready means for terrorists to both distribute their message and make fundraising appeals at little or no cost (Hinnen, 2004). For example, some Islamic news sites also include appeals for funds and directions on where to submit them (Dartmouth College, 2003); three charities that rely on Internet fundraising—the Benevolence International Foundation, the Global Relief Foundation, and the Al-Haramain Foundation—have had their assets frozen by U.S. authorities because of alleged ties to al Qaeda (Weimann, 2004).

The Future of Resource Acquisition

It is logical to expect that terrorist groups will continue to rely on modern financial transfer systems enabled by improved technology. As in the other aspects of terrorist operations that we have examined, network technology presents both opportunities and vulnerabilities for resource acquisition. For example, online purchasing offers terrorist groups a broader base of suppliers to support acquisition, although the use of such sites may increase the likelihood that security forces will detect efforts to acquire weapon components and other suspect material.

Another trend—the increasing sophistication of counterterrorism efforts, including reliance on enhanced legal authority in detecting and defeating the effective use of such tools—may actually reduce terrorist groups' ability to use some technology in the future. For example, in the period immediately following the September 11, 2001, attacks, a number of donors, charities, businesses, and informal or underground money transfer organizations had assets frozen or seized.[14] Despite improvements in transfer technology, such thinly disguised operations are often no longer viable due to increased law enforcement awareness, scrutiny, and legal authority. Nonetheless, as explained by the 9/11 Commission, completely cutting off financing to terrorist groups has been essentially impossible (Roth, Greenburg, and Wille, 2004), but the ability to close down revenue streams may not be the most effective course of action for security forces, since tracking financial flows has proven to be a very effective way to locate terrorist operatives and supporters and to disrupt terrorist plots.

Revolutionary improvements in terrorists' ability to acquire resources are most likely to occur in the areas of message distribution and funds transfer because they leverage one of the most important fundraising mechanisms for terrorist groups—contributions from support groups outside the country of operations.

In message distribution, terrorists will adopt modern advertising techniques used in legitimate businesses, such as sending tailored appeals directly to individuals who are likely to respond favorably. Tailoring the appeals to small groups or even individuals can substantially

---

banking system. Transfers leave no paper trail and offer anonymity to both the originator and the recipient. For an excellent discussion of how they operate and how they are used, see "Money-Transfer Systems" (2004).

[14] These included financial transactions by such entities as the Al Rashid Trust, a welfare organization that operated bakeries in Afghanistan, and Al Shamal Islamic Bank (established by Osama bin Laden in Sudan in the 1990s), with correspondent banks in London, Frankfurt, Geneva, and Johannesburg. See Wechsler (2001).

increase the efficiency of advertising messages, and we hypothesize that the same mechanism will generally apply to appeals for funding by terrorist groups.[15]

In funds transfer, technologies such as cyberpayments may enable the clandestine transfer of very large amounts of what we now think of as cash[16] quickly and securely and in a difficult-to-detect manner.

The small size of the devices used for cyberpayment tokens (credit card–size or smaller) in comparison to the bulk of physical currency that must be carried for large cash transfers, makes detecting the transfer of a cyberpayment token very difficult, if not impossible (Molander, Mussington, and Wilson, 1998). Further, if intelligence agents become adept at detecting electronic transfers from external message characteristics, terrorist organizations may even prefer the physical transfer of cyberpayment tokens to electronic transfer.[17] Whether or not these characteristics match the future needs of terrorist groups will depend largely on how cyberpayment systems are commercialized—in particular, the level of anonymity they guarantee.

The ability to secretly transfer very large amounts of cash allows terrorist organizations to buy influence on a large scale or to destabilize a local economy, which could directly support propaganda, influence, or recruitment activities in an area. Such procedures can potentially change how terrorist organizations use funding transfers. Presently, they are used to provide the funds necessary for operatives to conduct their attacks. With access to cyberpayment technology, money may become a tool to attack a local economy (posing legitimacy problems for the local government authorities) or to buy influence (supplanting or co-opting the local government authorities) to more directly achieve some terrorist goals. Such uses would probably be most effective in economically underdeveloped overseas areas, where terrorists are attempting to broaden their areas of control, than within the United States or other developed countries. However, transferring large amounts of money into a local community could potentially buy support and influence among disadvantaged populations within the United States as well.

Training

Depending on a group's sophistication and requirements, terrorist training may range from rudimentary lessons in the use of small arms and explosives to detailed instruction in advanced operational tactics and procedures, which can include the use of sophisticated technologies. Initial training of new recruits is often integrated with the indoctrination process.[18] Training at higher skill levels of either operational art or technical applications is more often conducted on an apprentice basis in the actual region of operations, unless the group has a safe haven or state sponsor through which such instruction may be developed and provided. Although clan-

---

[15] For a discussion of the strong basis for success in marketing, see Yuxin Chen, Narasimhan, Zhang (2001).

[16] That is money or purchasing power that can be used anonymously by any bearer.

[17] This is in contrast to detection by a means that requires an analysis of message content, which may be more difficult, as it involves defeating the cyberpayment system's encryption method.

[18] Discussion of a number of terrorist organizations' training regimens can be found in the case studies in Jackson, Baker, et al. (2005b).

destine groups that must operate without a safe haven are constrained, they have been able to train cell members for complex operational and technical activities by focusing on the tools and tactics most useful to them, such as bomb making and operational security.

Technology may complicate training; operating more sophisticated devices often requires members with more advanced skills; however, terrorists have lesser needs for sophisticated systems than do security forces. For example, U.S. and allied military forces have strong incentives to develop and use complex systems because they are expected, often for very good reasons, to counter adversaries using technology rather than personnel. Terrorists are largely free from such expectations and are likely to adopt technologies that require substantial training only if they promise an operational advantage over existing capabilities that are often adequate for conducting a terror campaign.[19]

Terrorist training has traditionally relied on technology in only limited ways. In some cases, sponsor states, such as Iran, Syria, and Libya, have facilitated technology-related training as part of their intelligence apparatus or paramilitary training programs.[20] Iran, for example, has provided such training to a variety of terrorist organizations, most consistently to Hizballah in Lebanon (Cragin, 2005), other sympathetic states have provided training to Palestinian terrorist organizations, and Libya has provided support and training to PIRA (Jackson, 2005). In most of these cases, training was provided through face-to-face interactions between state-provided experts and the groups, frequently in the supporting country. Any technology in use, such as voice recorders and communication devices, was largely limited to that used in military or intelligence establishments at the time.

Training by terrorist organizations themselves was similarly hands-on and seldom relied on technology to enable training beyond using the group's equipment as training aids.[21]

Current State-of-the-Art Training

The development of the Internet has led to what *The Washington Post* has termed "a massive and dynamic online library of training materials" in multiple languages that not only cover various weapons and attack strategies, but also provide instructions in traveling under cover and forging identities (Taylor, 2005; Coll and Glasser, 2005).

Video has recently become an important component of technology-enabled training. In the past five years, the production and use of video recordings in terrorist operations and training has increased substantially (Lamb, 2002). In Afghanistan, Iraq, and Chechnya, individuals not directly engaged in conducting attack operations regularly record many operations.[22] These recordings provide not only a resource for operational training, but also a number of

---

[19] Jackson has identified several factors that motivate terrorist groups to adopt new technologies. These factors include the technology's operational utility, the group leaders' risk averseness, and the organization's operational style. See Jackson (2001).

[20] See Byman (2005) or Hoffman (1998) for reviews of state sponsored training of terrorist groups.

[21] For example, descriptions of PIRA training activities reveal predominantly face-to-face instruction without mediating technologies ("Five Days in an IRA Training Camp," 1983; Collins and McGovern, 1998; O'Callaghan, 1999).

[22] For some examples of these engagement videos, see, for example, "Chechen Ambush" (2006), "Iraqi Improvised Explosive Device Attack" (undated), "Preparing and Employing a Landmine" (undated), or "Ambush in Afghanistan" (2007).

related products at the same time. These include material for propaganda, a source of damage assessment that allows the operation's success to be judged, and the intelligence on security force reaction that is necessary for developing new terrorist tactics. The ability to create several products simultaneously can increase the efficiency of both training and operations. For example, a well-produced video can be used as an attack-planning tool, as an after-action assessment of an attack's effectiveness, and as a tool for training less experienced team members. Video recording also makes it possible to present some technical skills, such as bomb placement, to trainees who do not read (or do not read the language used by the trainers). Even seasoned group members are likely to find that video recordings are far superior than recall based on human observation or other recording methods such as still photography as a means of identifying weaknesses in target security measures or patterns of security force movement that can be used either for operational planning or for training.[23]

Terrorists have also started to use a related form of network technology—computer simulations and their associated graphics—to train for missions. For example, Zacarias Moussaoui (Federal Bureau of Investigation, 2001), Mohammed Atta, and the other 9/11 hijackers had used Boeing 747-400 flight simulation software as well a Boeing 767 flight deck video as part of their preparation "to increase their familiarity with aircraft models and functions, and to highlight gaps in cabin security."[24] The use of such simulations enables a great deal more repeated practice and a much lower profile than would be possible by conducting practice missions in some actual setting.

The Future of Training

Given the substantial advantage that video recording provides for training as well as other aspects of the terrorist activity chain, we would expect to see such applications become increasingly prevalent. The decrease in size and cost of video recording equipment, coupled with other improvements such as ease of use, better optics (for longer-range observation), picture stabilization, and low-light–level recording will likely accelerate this trend to the point that videotaping an attack is as regular an operational task as movement or concealment. We expect that the primary payoffs of the pervasive use of video recording technology will be in the ability to train recruits for more complex tasks than would otherwise be the case and in the efficiency with which high-quality materials can be produced and distributed for multiple purposes. The latter point is particularly the case with the ongoing shift to digital video recording, which enables the use of the Internet to facilitate distribution.

In considering whether advances in network technology could promise revolutionary changes in terrorist training, perhaps the most compelling argument can be made for advanced computer-based games, including MMOGs. Computer-based gaming has been adopted on a

---

[23] PIRA reportedly used videotapes of "challenge and response studies" (i.e., sending a PIRA member who was known to the security forces into an area, then monitoring the reaction of any overt or covert surveillance officers to his appearance) to supplement their countersurveillance training (author interview with law enforcement official, Northern Ireland, May 2005).

[24] National Commission on Terrorist Attacks upon the United States (2004, p. 168). The hijackers reportedly also watched movies with hijacking scenes.

large scale by our national military forces; games that are sophisticated enough to provide a challenging and realistic training environment are now available for nearly every level of combat and command.[25] More correctly termed *learning games*, they deal with a range of operational and leadership challenges and provide the services with a sophisticated means of teaching problem-solving in a variety of situations. The range of skills that can be taught using these games runs from those needed to make decisions in combat situations (e.g., who to engage and who not to engage) to those needed to understand the opportunities and demands associated with joining the military. The latter skill involves a complex problem in individual decision-making that deals with a great many factors. The U.S. Army has successfully addressed this learning challenge with its well-known online game *America's Army* (see U.S. Army, undated). Games that teach leadership skills are now regularly used by the military in its professional officer education, as well as by business (see Stitt and Chappell, 2005; and Sawyer, undated). Although none of these learning tasks is directly applicable to terrorist training, the general classes of operational issues that they address may be well suited to the requirements of terrorist groups. These requirements include engagement decisions (which security forces to avoid or which target to engage), leadership skills, training individuals so they make the right decision with respect to recruitment, and matters associated with the interconnected tasks of indoctrination and basic recruit training. Although the evidence is lacking that such learning games so significantly enhance training that they can lead to revolutionary changes, the evidence that such approaches to training produce more capable team members is strong enough to have convinced both our military and many businesses to make notable investments. And there is a growing body of evidence that such games are particularly effective for individuals whose life experiences include online gaming for entertainment—the "digital generation" (Steinhuehler, undated).

Creating False Identities, Forgery, and Other Deception

The creation and use of false identities, forgeries, and other deceptive techniques can be useful across a broad spectrum of terrorist activities.[26] This may take the form of disguises donned by operatives or forged documents that may allow them to infiltrate a target facility; operatives may become employees and work at the targeted facility for some time to gain access to critical areas or information; and forged documents can allow personnel and materiel to be transported unhindered, even under the watchful eye of security forces.[27] Assuming false identities may take the form of repeated hoaxes (feints and demonstrations, in military parlance), which

---

[25] For a selection of these games, see DARWARS (undated).

[26] See Gerwehr and Glenn (2000, 2003) for a review of deception.

[27] Many terrorist organizations have adopted the use of deception in documents and identification to evade security and intelligence measures. See case studies in Jackson, Chalk, et al. (2007).

may greatly increase the assessment burden by misdirecting intelligence resources or reveal the response apparatus of the security forces.[28]

However, despite their potential utility, terrorist groups' deceptive capabilities have, until recently, been rather crude. Unless subsidized by a state sponsor, the quality of their instruments—disguises, forged documents, decoys, diversions, and disinformation—has generally not been particularly sophisticated. For example, during the 1950s, the Algerian Front de Libération Nationale (FLN) used well-dressed European-looking women to slip through checkpoints and carry out bombings. Similarly, the Movimiento Revolucionario Túpac Amaru (MRTA) of Peru slipped past security in its takeover of the Japanese embassy in late 1996, but the disguised vehicle (an ambulance) and forged identification were of only modest sophistication. A thorough check by security personnel should have uncovered the ruse (Nelan, 1997). Only groups engaged in multiyear sustained campaigns of terror (for example the PIRA, Jewish terrorist groups active in Palestine, and Sendero Luminoso) have appeared to contemplate sophisticated deception campaigns, such as repeated, concerted hoaxes to burden and confound the intelligence apparatus of their adversaries.

These patterns may reflect nothing more than a judicious economy of effort, since, with a few exceptions like the Green Zone in Baghdad, most terrorist targets are not protected by sophisticated security arrangements and, in fact, are chosen precisely because of their vulnerability.

Current State-of-the-Art of Deception

The advent of widely available network technology such as high-quality color printers and Web access along with improved image manipulation software has greatly expanded the range and effectiveness of deception options available to terrorist groups. This is primarily a result of two factors. First, *knowledge* of how to exploit the weaknesses of adversaries may be made globally available; it is no longer exclusively held in the heads of a small number of individuals. For example, at the time of the first battle of Grozny (1995), Chechens who had previously been members of the Russian armed forces needed to make firsthand contact with their comrades to inform them of Russian security methods, such as communication protocols, which could allow the Chechens to impersonate Russian units. But a mere 10 years later, when Chechens discovered weaknesses in Russian tactics and procedures, that information was posted on a Web site and became available across the entire globe. In another example, it is now commonplace for techniques for forging documents, disguising the operative, and camouflaging observation positions or vehicles to be compiled and disseminated among jihadist Web sites and online forums.[29]

Second, the *means* of using knowledge for deceptive ends have become widely available. In the recent past, most terrorist groups lacked the technical skill and resources to doctor a photo's content; even those who could muster such capabilities were typically able to produce

---

[28] For example, PIRA engaged in repeated deception operations targeted against the security force's telephone tip lines in an effort to confuse the police and to lead officers into assassination attempts and booby traps (Jackson, Chalk, et al., 2007).

[29] See, for example, the extensive collection available from the SITE Institute (undated).

false passports or other forged identity documents in only one or two locations dedicated to that activity. For example, in 1972, the Baader-Meinhof group maintained a facility in a rented Hamburg apartment exclusively for creating forged documents and disguises.

However, the global spread of inexpensive, powerful personal computers, scanners, photo-editing software, and printers now allows the terrorist groups to produce authentic-looking forged documents and identity photos almost anywhere. Most documents and images produced in this fashion will usually not withstand a detailed forensic analysis, but they may be made good enough to withstand cursory inspection by an undertrained or hurried clerk, security guard, or police officer. For example, the terrorist organizations that stormed the Indian Parliament on December 13, 2001, apparently gained access to the building with forged passes and other documents (Mishra, 2003). The operatives who assassinated Ahmed Shah Masood, the leader of the Northern Alliance resistance to the Taliban government in Afghanistan, also used forged identity documents (Davis, 2002). Although other groups from the Liberation Tigers of Tamil Eelam (LTTE) to the Real Irish Republican Army (RIRA) to al Qaeda have demonstrated the ability to produce high-quality forged documents or disguises, these skills need not necessarily be part of the core competencies of the terrorist group itself, as the technology also enables the production of such documents to be outsourced to other organizations, both legal and extralegal, or to sponsor states.

The Future of Forgery and Other Deception

Two further technological developments could result in a significant gain in terrorist capabilities for deception, both of which may be fairly characterized as logical extensions of current trends. The first development would involve terrorists developing the capacity to determine the exact parameters of critical public or private identification documents (or devices as in the case of radio frequency identification [RFID] chips or smart cards). If these parameters were known, the new generation of identity documents, which are designed to incorporate a high degree of veracity and authenticity, could be forged. Coupled with a second development, the mass distribution of duplication or manufacturing capabilities based on these parameters, this could substantially complicate the problem facing security forces because of the ability to mass produce very high-quality forged identity documents.

However, this vulnerability results primarily if advanced technologies are used to counterfeit traditional identity documents. If advanced network technologies were used as the basis for a future identity system, identity documents could become more reliable than they are presently. For example, electronically read identification cards can include a digitally signed bit string composed of a photograph or some other biometric; the identification name or number; and information about the date, time, and location of the card's issuance and registration in the identity system. Ensuring that a card is a valid identity document would involve electronically reading the information on the card, including the photo or biometric, and ensuring that the digital signature that was computed from that information and the private key used in issuing the card was the same as the digital signature on the card. Because it is virtually impossible to generate the correct digital signature of such a card without a private key, a falsified identity card that had not been issued through the official process could not be validated by the system unless it were an *exact* copy of an existing identification card. Any attempt to alter the digitized information (including the photo or other biometric) that is part of the digital signature will prevent the system from validating the card. This feature could provide much stronger protection against the now-common practice of forgery, but it would have some vulnerabilities. First, if the issuance and registration process were not secure, insiders could issue cards that would be recognized as valid. Also, a person who looked sufficiently like the person in the photograph on the card could use an exact copy of a valid card (although, with other biometrics, such impersonation might be much more difficult). Finally, if identification checks were only done by visual inspection rather than electronic reading because readers were not available or broken, the system would be no better than today's.

Although existing trends in terrorist groups' deception capabilities may be becoming more problematic, the potential for further, and perhaps revolutionary, development in terrorists' deception skills exists. One such advance would be the ability to impersonate, without detection, any person engaged in electronically mediated communications, such as videoconferences, phone calls, or simpler communications such as emails or text messages.

The world increasingly relies on electronically mediated communications for everything from finance to news reporting to military orders. Electronic communication provides only a limited set of cues that might alert a reader that the author of an email is other than who he or she claims to be, and impersonations are commonplace in email traffic. Tools for verifying authorship of documents, for example, are available in the form of stylometry[30] and other techniques, but, if algorithms are developed that can defeat those methods, impersonations could be carried out without detection until better methods of detecting deceptions become available.

The consequences of these impersonations (e.g., spoofed air traffic control communications, false military orders, phony financial transactions) could be extremely harmful. Other forms of electronically mediated communication such as video teleconferencing may also be vulnerable to deception techniques, since such video may be doctored or even completely forged from start to finish. With a sufficient corpus of recordings on an individual, totally falsified videos may be generated with the individual saying or doing virtually anything the forger wishes. Although seemingly the stuff of science fiction, this capability has already been demonstrated (Emery, 2002). Researchers at the Massachusetts Institute of Technology (MIT) have combined artificial intelligence and videography to make words appear to emerge from the lips of public figures who could never have said them. Examples include Marilyn Monroe, a movie star of the 1950s and, more recently, a pop icon, lip-synching a song that was not written until decades after her death, and Ted Koppel, ABC's *Nightline* anchor, speaking in Spanish.

---

[30] Stylometry is the use of statistical analysis of style and word usage in texts over time to detect changes that might indicate impersonation or deception.

Reconnaissance and Surveillance

Reconnaissance and surveillance are critical functions for any terrorist group; the success or failure and the impact of any attack frequently depend directly on the quality of the reconnaissance and surveillance that preceded it. For example, the destructive LTTE July 2001 attacks on Sri Lanka's Bandaranaike airport were, in large part, so damaging because of the terrorists' detailed knowledge of the airport's layout.

Historically, effective terrorist groups invest time and effort in both reconnaissance and surveillance through a wide range of activities over periods that may extend for years. For instance, PIRA, the LTTE, and al Qaeda have performed surveillance of their intended targets for years before striking.[31] Of course, there are examples of attacks following much briefer reconnaissance and surveillance periods. Members of Loyalist groups in Northern Ireland indicated they ran operations without any preparation at all.[32]

In the past, terrorist groups have performed reconnaissance and surveillance with low-tech methods, and the information collected during such activities has been used in relatively simple planning. For example, most historical terrorist reconnaissance and surveillance have consisted of operatives physically scouting a target site—examining its perimeter, its traffic flow, its guard patrols, and the like. Such surveillance also entailed creating hand-written descriptions and drawings of the facility and locating the facility on a map (*United States v. Usama Bin Laden*, S[7] 98 Cr. 1023 [LBS] S.D.N.Y., February 21, 2001, pp. 1190–1192, 1142–1147).

Operatives have used still photos of a target, which were then developed in a group operative's darkroom or by a private company. At its training camps in Afghanistan in the early 1990s, al Qaeda provided instruction in using cameras and how to develop photographs without being detected (*United States v. Usama bin Laden*, S[7] 98 Cr. 1023 [LBS] S.D.N.Y., February 6, 2001, pp. 1142–1147). In late 1995, for example, al Qaeda operatives set up a darkroom in an apartment in Nairobi, Kenya, to process surveillance photos of the U.S. embassy there (*United States v. Usama bin Laden*, S[7] 98 Cr. 1023 [LBS] S.D.N.Y., February 21, 2001, pp. 1190–1192).

Manual reconnaissance methods are error-prone and biased in exactly the same ways as the scientific literature on eyewitness recall demonstrates.[33] For example, J. Bowyer Bell, an insurgency expert who spent several decades studying PIRA, has frequently noted the high ratio of aborts to attacks by PIRA operatives; these aborts occurred because faulty intelligence had been gathered, which became apparent only when the operation was about to begin (Bell, 1998, pp. 450, 470). Because of such limitations in the past, few groups could afford to separate—either operationally or logistically—reconnaissance and surveillance operatives from those executing actual attacks. However, it is useful to note that, although reconnais-

---

[31] The extensive planning often conducted for terrorist operations are highlighted in two reports: National Commission on Terrorist Attacks upon the United States (2004) and Singapore Ministry of Home Affairs (2003).

[32] Author interview with law enforcement official, Northern Ireland, May 2005.

[33] See, for example, Heaton-Armstrong, Shepard, and Wolchover (1999) for a review.

sance methods such as video or photography remove some sources of errors, they can introduce sometimes-subtle errors in interpretation that can take some time to eliminate.

Current State-of-the-Art Reconnaissance and Surveillance

Although current network technologies have changed reconnaissance and surveillance by terrorist groups, they have not obviated the need for terrorist operatives to scout a target physically; indeed, this remains the prevailing procedure for terrorist reconnaissance and surveillance. However, many innovations in network technologies have greatly enhanced terrorist operatives' ability to perform reconnaissance and surveillance tasks. Ubiquitous and inexpensive digital cameras now allow for video to be recorded in addition to individual still photos. Such images do not need to be developed, eliminating the need for equipment that might attract attention or the need to rely on outsiders for this process. Further, the images can be readily edited with inexpensive desktop software and hardware. In preparation for a planned attack at the Yishun Mass Rapid Transit Authority station, a mass transit point in Singapore that U.S. military personnel and their families frequented, Jemaah Islamiyah supplemented its video surveillance with an explanatory voiceover. Although combined use of audio and video demonstrates an advance in surveillance and reconnaissance capabilities, it also shows vulnerabilities that new technologies can create: The operation was compromised when a copy of the video was found in Afghanistan during Operation Enduring Freedom (Baker, 2005, p. 81).

Digital photography is only one of the current resources available through network technology to terrorists conducting reconnaissance and surveillance. More adaptive terrorist groups can now have ready access to the reconnaissance and surveillance value of commercial satellite photography, GPS, and extensive data about potential targets available on the Web; such data can be perused and obtained anonymously with relative ease. These techniques can move reconnaissance and surveillance to increased standoff ranges, lessening the chance that operatives may be detected. Operatives conducting reconnaissance and surveillance can also transmit findings almost instantly to operatives or planners or logisticians in another part of the world with relative ease, and, with appropriate operational security (OPSEC) and encryption, the risk of exposing themselves or their activities to detection is modest.[34] Finally, such groups as al Qaeda have exhorted their supporters to become more involved in reconnaissance and surveillance on behalf of the group and transmitting their findings via the Web to those who are planning and executing operations.[35]

The Future of Reconnaissance and Surveillance

Two potentially troublesome future developments in capabilities for terrorist reconnaissance and surveillance are essentially logical enhancements in current technology trends. One is

---

[34] The technique of engaging home-grown operatives with a lower profile to use video recording to reconnoiter targets and sending it to terrorist planners operating in a safer environment was apparently the modus operandi of the Canadian and Miami groups disrupted by Canadian authorities and the FBI in mid-2006. See, for example, *United States v. Batiste, Abraham, Phanor, Herrera, Augustin, Lemorin, and Augustine* (S.D. Fla.) June 22, 2006, pursuant to activities in or about November 2005.

[35] For examples, see the numerous communiqués to this effect translated and published by the SITE Institute (undated).

that current trends toward miniaturization, standoff, and lower costs in the technology used in reconnaissance and surveillance are likely to continue. This trend makes it easier to obtain equipment that is capable of collecting information from greater standoff distances.

Second, the accuracy of reconnaissance and surveillance data may increase, thus increasing its value to terrorist groups, particularly if ways of collecting information are coupled with effective ways to pass that information to planners. For example, if digital photography were to include a feature that permitted the simple creation of mensurated[36] images without requiring expert knowledge for image analysis, terrorist planners would be better able to exploit such imagery for operational purposes. Such capability could inform activities such as entering a target area or delivering ordnance using guided weapons. If such information could be passed to planning tools in a way that required minimal expertise, more sophisticated planning may be possible.[37] For example, if digital imagery could be passed directly to a computer-aided design or modeling application that produced accurate three-dimensional images of a target site, sophisticated groups might exploit such precision in planning attacks. Cheap and readily available GPS technology can provide more accurate physical locations. Despite these advances, in the research team's judgment, it seems unlikely that such capabilities would completely obviate the need for physical observation of the target by experienced terrorist groups because operational nuances and physical details, for example the existence of a drainage ditch just behind a fence that a group might plan to breach with a vehicle, can pose significant problems during an operation. Network technologies, however, may well serve to limit the exposure associated with conducting such final reconnaissance of a target by focusing operational experts on those aspects of the target that require their skill and understanding.

Planning and Targeting

Although some form of planning takes place throughout the entire activity chain, we focus on the planning directly related to the operational aspects of an attack, especially target selection. Planning involves deciding on an operation's objective. Recent terrorist operations indicate that the strategic objectives of terrorist attacks are relatively constant; these appear to be inflicting mass causalities, causing economic damage, or damaging iconic targets.

Leaders of international terrorist organizations such as al Qaeda and its affiliates have shown that they understand the potential consequences of carefully planned attacks on important and iconic targets as well as the psychological value of demonstrating that they can strike

---

[36] A mensurated image, which allows for an object's true location and dimensions to be extracted, can be very useful in some types of attack planning activities, as well as during the attack. The utility of such information to terrorists depends, of course, on the method of the attack.

[37] Current network technology already supports such techniques for some purposes. When Abu Musab al Zarqawi left behind communication gear in an escape from U.S. troops in Iraq in about April 2005, the U.S. military found a catch of keychain computer drives, devices that can be easily hidden and passed by handshakes. These are reputed to have become an al Qaeda trademark (Windrem, 2005).

in the U.S. homeland. Attacks that exploit U.S. infrastructure as weapons against itself[38] allow terrorist groups to leverage their capabilities and cause destruction beyond the level that the groups could support logistically with their own assets.[39] Such attacks involve an additional overt action, seizing or gaining control of the infrastructure asset. Although this eases logistical problems involved with weapons, it does add complexity to the planning.

Historically, targeting and target identification involved collecting information through local first-hand knowledge of potential targets, news media outlets, and personal communication. More recently, terrorists have used other media outlets such as the Internet and digital media sources to gather information to support selection of specific targets. In the past, terrorist groups have typically constrained themselves to stationary targets (e.g., defined areas, buildings) or individuals at a given location or ambushes in which the target passed by a predetermined location—all designed to lower uncertainties that the group faced during the operation.

Current State-of-the-Art Planning and Targeting

Planning has changed as a consequence of technology, particularly connectivity, mobile computing, and IT services. The Internet has made research for operations and planning easier and allowed some ability to adapt to target movement or response. The U.S. government's indictment of Zacarias Moussaoui included reference to a laptop with a flight simulator, pilot procedures for a Boeing 747, and information on crop-dusting ("Ridge Wants Tech Firms to Enlist in Terrorism Fight," 2002; Dartmouth College, 2003). A laptop belonging to al Qaeda in Afghanistan reportedly contained design details about a dam and engineering software, suggesting that the organization was studying ways to attack such facilities (Gellman, 2002).

Video reconnaissance, typically developed under the guise of tourist or other innocuous activities, has helped streamline terrorist planning processes. As noted already, the use of video makes it easy to pass information directly to planners.[40] Other improvements include the digitization of terrorist planning manuals such as the *al Qaeda Training Manual* (U.S. Department of Justice, undated). This allows these documents to be used on laptop computers, easily transferred via compact discs, and posted on the Internet. Some useful planning information may be acquired from the Web, but recent studies indicate that, in most instances, it is not of sufficient resolution or reliability for terrorists to use it in final planning because of the risk from flawed or incomplete data to an operation's success (Baker et al., 2004). It may, however, allow groups to focus their physical observations and thus lower the amount of exposure associated with reconnaissance.

---

[38] For example, using a tankship or tankbarge with a flammable or explosive cargo (such as liquefied petroleum gas or liquefied natural gas) as a firebomb against a port city that has densely populated areas close to shipping channels; using aircraft as missiles as was done in the September 2001 attacks; or releasing volatile toxic chemicals (such as liquid chlorine or anhydrous ammonia) from a chemical plant or a vessel near a populated area.

[39] In the ship example, an equivalent amount of explosive or toxic chemical would be very difficult for a terrorist group to acquire or transport into the United States.

[40] See, for example, discussion of Jemaah Islamiyah planning processes in Baker (2005).

The Future of Planning and Targeting

With the ready access to encryption that gives private individuals a high degree of confidence that they can communicate securely with their peers, we would expect to see a greater use of digitized video and photography being used directly by planners to evaluate potential targets. Such technologies could help to reduce the risks that operational plans and group members' identities will be revealed if authorities seize planning documents, as was the case with the Jemaah Islamiyah planning footage found after the invasion of Afghanistan (Baker, 2005).

Access to a wide variety of indexed and searchable sources of real-time images (generated by real-time webcams or video playback systems, which have become increasingly common) may enable terrorist groups to search target information the way static images are currently collected for target selection. The ability to find relevant video footage and use video cameras to monitor targets in real time may allow terrorists to better determine the times and circumstances under which the targets are most attractive, such as peak ridership times on transportation systems, when and where crowds gather in public places, how train schedules coincide for coordinated targeting and attack, and even how and when emergency responders react to an incident so they may be targeted in follow-on attacks. The increasing availability of information on the Internet may also enable terrorists to consider a wider range of possible targets during the initial target selection process. Terrorists may learn more about the significance of possible targets, allowing them to select higher-value targets, be they collections of people, symbolic targets (e.g., monuments or tourist sites), or infrastructures (e.g., power stations or financial data warehouses).

With the use of more sophisticated information-reliant fuzing devices (beyond current use of cell phones for that function) that allow existing explosives to be used at a time, at a place, or in response to a sensor so the attack more effectively matches the vulnerabilities of targets, terrorists may select targets that heretofore could not be attacked effectively, such as mobile or fleeting targets or security forces and first responders reacting to an initial attack.

Revolutionary improvements in planning could change the currently very slow and drawn-out operational tempo of terrorist attacks. Although the popular press casts this as a strength—part of al Qaeda's patient plan—if a terrorist organization could mount attacks on New York, Madrid, and London in the same week or month, the effect would be substantially greater than that of historic attacks, if only in its effect on the terrorists' support and recruitment base populations. Such groups may be patient, but this is most likely the result of the need for care in planning and mounting an attack, not because such groups believe that slowly disrupting the West is the most effective strategy.

To do this, network technology would need to offer the possibility of planning operations with the speed demonstrated by the best modern militaries. Building this capacity would probably require simulation and modeling and secure communication, as well as collaboration and decision tools. Although such technologies are becoming increasingly available commercially, they must be complemented by high-quality data and improvements in other operational and support capabilities. Further, effective rapid planning requires well-trained and experienced planning experts. Thus, although network technologies that permit rapid planning could enable a revolution in the pace of operations, they would probably not be sufficient in themselves. As is often the case with technology-driven improvements in any process, several different technologies and sufficient people with the right expertise must come together for the greatest increases in effectiveness.

In selecting targets, terrorist groups typically prefer to emphasize the most iconic, casualty-prone, or economically significant assets and infrastructure targets (U.S. Department of Homeland Security, 2005). Given the visibility of such targets in Western society and their general vulnerability, terrorists already have such a "target-rich" environment that it is difficult to conceive of any technological change that would significantly increase their ability to identify attractive or vulnerable structures.[41] But innovations in network technology that allow terrorists to do things that are currently impractical could open up different, and potentially revolutionary, options. This might be particularly likely if it allows the use of weapons or tactics that could have markedly different effects than attacks to date.

For example, the uncertainties associated with the effect of releasing chemical and biological weapons (e.g., weather conditions that might significantly influence operational outcomes) have been cited as a disincentive that keeps risk-averse terrorist groups from pursuing these agents (Donahue, 1999, p. 22). The use of plume modeling or infectious disease simulation software could possibly make attacks with chemical or biological weapons more practical by allowing preplanned evaluations of how an agent is likely to affect the target or assessments of current conditions to indicate when they are conducive to an effective attack.[42] Network technology tools that allow terrorists to experiment with operational scenarios that are currently seen as too uncertain to justify the investment of scarce resources could also raise confidence in making significant shifts in the use of weapons and attack tactics.

---

[41] See Baker et al. (2004) for a discussion of the level of public accessibility of information about potential target sites.

[42] The utility to terrorists of plume modeling software, or the danger from its availability, is a complex question. Plume modeling software is available from both commercial and noncommercial sources and, in some cases, on a free trial basis (for examples, see Environmental Health Safety (2006) for a collection of 43 plume and particulate transportation models for direct downloads or for links to the owning entity's Web site for models such as CAMEO® (U.S. Environmental Protection Agency, 2006), and HYSPLIT (U.S. Department of Commerce, undated). These tools are typically used in the environmental, scientific, incident management, and military communities to model the impacts of dangerous emissions from either unintentional or intentional releases. These tools, when combined with geographic information systems (GISs), are useful contingency planning and decision-support tools, especially for first responders and incident managers who are working from current meteorological data in response to incidents in which chemical, radiological, or biological material may have been released. In these applications, understanding where sure-safe areas are is very important, and, consequently, the models place their greatest emphasis on the plume's edge, which defines the area where at least some hazard might exist. In contrast, attackers are usually concerned about performance in the core of the plume where high confidence attacks can occur and on the predictability of a planned attack. In many cases, it is not possible to confidently choose the parameters that have a great influence on model output, let alone have confidence that the model output can give useful predictions about a specific attack. A serious user of these tools would likely discover the large uncertainties associated with modeling a particular attack and the difficulties associated with eliminating those uncertainties. As a result, they would probably choose to launch their attack in a manner or location that did not depend on predictions from detailed modeling (thus obviating models' usefulness). Monitoring individuals who download, use, or ask questions about the models might produce a payoff to security forces if it identifies individuals with a suspicious set of interests, but this would require close monitoring by a person with a good knowledge of the models, the user community, and their typical uses. Although this is possible, it is also resource intensive.

Communication

Instead of trying to enumerate all of the different communication technologies and how they affect the other terrorist functionalities, this section focuses on two cross-cutting aspects of communication that affect all of the functionalities we have identified in the activity chain: the security of communications, including the growing importance of encryption, and the strengths and vulnerabilities of the growing number of different modes of communication.

Security in communication relies on several fundamental techniques: translating messages into a hard-to-interpret format (such as encrypted messages or the use of code words) that conceals their true content, precluding unintended parties from intercepting the communiqué, and seeking anonymity in communication to mask the role of the message or to avoid compromising the sender.

Current State-of-the-Art Communication Practices

Historically, operatives have sought secure communication through techniques such as changing call locations and times, keeping communications brief, disguising their voices, using code words, and endeavoring to achieve anonymity through the use of prepaid phone cards or stolen phones.[43]

Although terrorists have long had access to ciphers, currently other techniques are often used to achieve the same ends without the burden of using a cryptographically secure communication systems or materials that might in themselves attract attention. Much of the planning for September 11, 2001, was done using unencrypted email messages, albeit often written using simple codes sent from multiple locations using various ISPs to complicate the problem for U.S. security services (Campbell, 2001). The hijackers were not detected, in part, because of their use of tradecraft and their use of a communication medium that carries huge amounts of traffic every day.

Such operational measures have more recently been augmented through the use of computer-based cipher systems, which have become increasingly practical because of numerous applications that have been developed and continue to be developed to provide encrypted communication. Denning (1999) provides a good overview of the advent of the use of encryption by terrorist and criminal organizations in the late 1990s (Denning and Baugh, 1997). She points out that, as criminal and terrorist organizations adopt information technologies that permit them to communicate and store new kinds of data, they also seek to protect that data with encryption technologies. Denning's primary observation is that law enforcement and intelligence services need to expect to encounter data encrypted using a combination of home-grown and commercial encryption systems that will, in many cases, slow the pace of investigation absent some preestablished mechanism for defeating the encryption.

The ongoing proliferation of modes that use *digital* technologies enables low-cost encryption that can greatly complicate the job for the security services searching through the messages. The problem for security forces is that, unless the cipher (the algorithm for encrypting

---

[43] See the discussion of the communication practices used by Jemaah Islamiyah for insights into the use and limitations of such canonical techniques in Jackson, Baker, et al., (2005b, pp. 75–76).

the message) is susceptible to rudimentary exploitation techniques that can be exercised with little analyst involvement, a message can take substantial time and resources to decrypt, and it is not possible to predict whether the cipher is vulnerable to exploitation or whether the message is valuable.

These limitations, plus the substantial public policy issues they raise, imply that encryption can importantly disrupt intercept operations by security forces.[44] Nonetheless, a dedicated code-breaking effort by security services with adequate resources and time can overcome many forms of encryption. In the end, the encryption's value is that it buys time for the terrorist organization. This results in two key considerations for terrorist organizations considering the use of current encryption capabilities: (1) How long does the message need to remain secret to allow the terrorist organization's plans to be successful? And (2) is it possible to determine when communications have been compromised? Terrorist operations that require multiyear preparations thus put a heavy burden on the technique employed by the terrorist, making it likely that a multilevel scheme for protection—combining encryption with the use of other mechanisms such as code words—will be employed.[45] Such a multilevel approach to communication appears to have been used during the planning for the 9/11 attacks; Weimann (2004) reports that "thousands of encrypted messages" were posted in a password-protected area of a Web site in the run-up to the attacks and that the conspirators used public Internet access points and free email services to maintain their anonymity.

Another approach to making the message difficult to read is to hide a covert message within an overt communication. There has been considerable speculation about the potential use of steganography—the process of hiding data in other media formats such as images and music files (Kelley, 2001). Steganography is potentially significant because it can increase the layers of protection afforded sensitive information; having an effective code-breaking method is of little use if one cannot detect the encrypted message's presence in the first place. However, a SysAdmin, Audit, Network, Security (SANS) Institute assessment in 2003 found that there had still been no credible evidence that terrorists have used steganography or other watermarking[46] techniques to hide information in images or other formats (Lau, 2003). Although steganography does not appear to be a current or near-term capability, it may be prudent to monitor the technology and its progress because of the impact that it could have on security.

The second fundamental technique that can be used for secure communication, avoiding message interception, can also rely on a number of mechanisms. Of particular interest currently is the growing number of modes of communication that the market offer to consumers and terrorists alike. Any particular mode of electronic communication can be described as consisting of three components:

---

[44] For a discussion of some of the issues with key-escrow systems, see Chapter Five of Dam and Lin (1996).

[45] For an excellent discussion of what a terrorist might have to consider as a plausible set of attacks against strong encryption (such as pretty good privacy, or PGP) see Schneier (2000, pp. 324–333).

[46] In digital media, the watermark is a pattern of bits inserted into a digital image, audio file, or video file that is difficult to detect upon casual inspection and that can contain information that is used to convey a hidden message. Watermarks often are used to record a file's copyright information.

- the electronic device (e.g., cell phone, computer)
- the application (e.g., email, text messaging, VOIP)
- the communication technology used to pass the information (i.e., wireless fidelity [WiFi] network, code division multiple access [CDMA] cell phone network, wired Internet).

Changing any of these components can pose problems for security forces if they are not well equipped or technically capable.

Terrorists typically communicated throughout the 1990s with single-mode equipment, which included facsimile, fixed-line telephones, and some mobile telephones, despite their characterization in the al Qaeda training manual (U.S. Department of Justice, undated) as having only "modest capabilities" for security. In Sudan, al Qaeda operatives used two-way radios procured from the Sudanese Army because the radios were seen as more secure than telephones (*United States v. Usama bin Laden*, S[7] 98 Cr. 1023 [LBS] S.D.N.Y., February 6, 2001, p. 308; *United States v. Usama bin Laden*, S[7] 98 Cr. 1023 [LBS] S.D.N.Y., February 13, 2001, pp. 454–455), and, by 1995, top al Qaeda leadership had access to satellite phones (*United States v. Usama bin Laden*, S[7]98 Cr. 1023, S.D.N.Y., February 21, 2001). All these systems required cumbersome coordination and manually changing modes of communication as the sender and receiver had to coordinate the physical change from one type of equipment to another if they wanted to use multiple modes of communication. This complicated process limited the technique's practicality. To lower the chance of intercept, operatives primarily relied on changing call locations and times and keeping their messages brief.

Over the past decade, however, the modes of communication available to both general users and terrorists have proliferated and become more flexible, and a single device may now offer several modes in one package. The modes of communication that are now more readily available to terrorists that have been documented include cell and satellite phones of varied types, email, instant messaging, short message services (SMSs), Internet chat rooms, and weblogs.[47] More recently, the accessibility and affordability of services such as mobile telephony have improved markedly, even in very poor and remote areas.[48]

This growth in the various modes of communication and terrorist awareness of the utility of using multiple modes in electronic communication can present a serious challenge for security services that have not previously confronted terrorists with such high levels of technical expertise and operational acumen. For example, terrorists are likely aware that, although SMS and chat room messages might be thought of as similar, they actually pose problems for ill-equipped security services. Because of their particular combinations of application and communications, they require different intercept techniques and equipment than that necessary to monitor conventional cell phones carrying voice messages or computers with email.[49] Terrorists probably also recognize that those same technologies can assist well-equipped security services as they did in the aftermath of the July 2005 London bombings ("All Over Bar the Shouting?" 2005). Experienced terrorist organizations have shown an acute sensitivity to the vulnerabili-

---

[47] See, for example, Zanini (1999).

[48] See, for example, "Leaders" (2005) and "Business" (2005).

[49] See, for example, "Internet Makes Drug Traffickers Hard to Catch" (2004).

ties of their communication and the flexibility to take extreme measures in switching modes to help avoid such vulnerability when necessary. For example, in response to the threat of Israeli monitoring, a number of Palestinian terrorist organizations have banned the use of cellular telephones during Israeli military operations (Jackson, Chalk, et al., 2007). Similarly, Osama bin Laden reportedly relied on personal couriers instead of satellite phones as the hunt for him escalated in late 2001.

The final fundamental technique for communication security, ensuring anonymity, is also currently changing due to advances in network technologies. A number of tools that permit anonymous communication on the Internet are currently available or under development to enable dissenters in countries with oppressive governments to exchange and distribute information without fear of identification or reprisal (Goldberg, undated). These tools are expressly designed to protect the communicating parties' identities as well as to protect their messages' contents.

Future Communication Technologies

Terrorists' use of encrypted communication can be expected to become a fact of life. In particular, it seems likely that the advances in cell phone computational capabilities will lead to easy-to-use encryption and decryption being used for all cell phone calls. The development of such ubiquitous strong encryption combined with technology to avoid tracing of communication produces the potential for enabling secure anonymous communication for planning, staging, and executing terrorist activities (Sui et al., 2004; Goldberg, Wagner, and Brewer, 1997).

Integrated encryption tools and downloadable privacy software that allows users to encrypt personal phone calls, text messages, and other communications among their electronic devices are likely to become widely available. Currently, VOIP providers such as Skype® can provide users with integrated encrypted voice communication over packet-switched networks without the need for complicated procedures on the part of the user. It is likely that, in the future, even rudimentary cell phone services will provide high-quality encryption in a consumer-friendly manner as a matter of routine service.

Growth in the use of virtual private network (VPN) technology may increasingly enable distributed networks to provide secure communication capabilities for organizations distributed across the Internet, and will form a backdrop of encrypted communication from which the terrorist communication must be extracted. Companies and organizations already widely use this type of communication technology to create local networks in remote locations. This technology is likely to become increasingly integrated into consumer devices and electronics, further increasing the background traffic of encrypted communication. In addition, consumer preferences seems to be continuing the trend to VPNs that are much less complex in their implementation, allowing more average individuals to establish private networks and share trusted, encrypted information.

From security forces' perspective, the availability of ubiquitous secure communication technology may prove to be less consequential than the associated reliance on the networks themselves. Advanced networks may attract clandestine groups because they offer secure communications. But secure communication may not lead to secure operations because even simple traffic analytic techniques can be useful in determining patterns over time and can allow a competent security force to develop a useful model of how potential opponents are organized or to even determine whether there is an organization in existence. Similarly, increasingly complex communication hardware and software leave open the possibility for designed-in exploitation by security forces.

From the terrorists' perspective, relying on network technologies for encryption may simply substitute a different set of vulnerabilities (the possibility of compromise through network exploitation) for the current vulnerabilities (the possibility that security forces may read nonencrypted communication). To take advantage of encryption to protect against the latter may require even greater reliance on network services, network technologies, and network providers, which may offer security forces a better ability to use traffic analysis or other advantages. Although this would ensure that the future would be different, it does not inherently advantage either terrorists or security forces, because such advantage usually results from the ability to capitalize on technical or operational errors whenever they are made.

In addition to ubiquitous encryption, future communication technologies may also provide the ability for seamless and dynamic shifts of communication modes from short-range wireless communication standards (e.g., WiFi, Bluetooth®) to conventional wireless cell phone frequencies (e.g., CDMA, global system for mobile communication [GSM]) and back. For security forces monitoring terrorist communication, such mode nimbleness can increase the challenges of successfully using terrorist communication traffic. For instance, a cell phone with only CDMA has only a single mode of communication, but a wireless cell phone with built in WiFi capabilities has two modes. When the device is in range of a WiFi network, the device can be used to make VOIP calls using the wireless connection to the Internet, and, when it is out of the range of WiFi networks, it automatically reconfigures itself to use the CDMA cell phone network. Terrorist access to easy-to-use devices with multiple modes of communication present challenges for security forces attempting to intercept or track communications because they must have the equipment and, in some cases, the legal authority appropriate for each mode. Since obtaining the proper equipment and authorities can rely on completely different infrastructures, technologies, hardware, and communication protocols, this can be a very complex and expensive undertaking.

One other future network technology that may offer revolutionary changes in capabilities for terrorists is autonomous networking. Portable handheld devices (e.g., PDAs and cell phones) have demonstrated the ability to use short-range wireless protocols (i.e., 802.11) to automatically set up local networks and to exchange data among like devices, independent of any other communication modes or infrastructure.[50] This eliminates the need for intermediaries such as switches and their support organizations (along with any potential linkage to security forces) to set up the communication network.

With such technologies, a network's emergence and growth can be described as *viral*—a genuinely new, bottom-up network formation process in which users (each of whom acts as both node and server) establish a network themselves, and the entire network self-assembles based on protocols. In most network arrangements, adding users increases interference and

---

[50] A simple example of such a system would be walkie-talkies operating on a discrete frequency known only to the handset users. Such a communication network does not require any intermediary equipment or organization outside the users.

degrades service. In this topology, network performance can increase as the number of users increases. Such communication networks can bypass security forces' centralized monitoring at switches or through the intermediate organization that manages the network infrastructure, and thereby provide fairly secure communication absent collection assets directed at the wireless traffic itself.

Future Communication Practices and Terrorist Activities

The trends toward pervasive encryption in handheld communication devices and access to multiple modes of communication create the possibility that terrorists may enjoy secure, difficult-to-trace, uninterruptible communication between individuals, no matter where they are. At first look, this possibility appears alarming. However, terrorist attacks seldom require orchestrating many operatives.[51] And because most attack planning and execution requires very low bandwidth and operatives generally move into the target area during the final stages of preparation, terrorists make very limited demands on communication in planning and conducting attacks. This is particularly true when suicide attacks are involved, because such tactics have even more limited communication requirements. As a result, terrorists already have a number of relatively secure, low-tech means at their disposal that are adequate for their limited needs in planning and executing attacks, so highly secure communication based on technology is not likely to tip the balance in ways that would be revolutionary. They may, however, limit the duration and impact of the successes that security forces are currently able to achieve.

In contrast to their effect on planning and operations, advanced network technologies for communication may result in substantial changes that benefit terrorist recruitment and propaganda efforts. This is because these activities require communicating with great numbers of people; therefore, these activities may benefit from podcast or multicast technologies because they offer greater leverage in terms of anonymity, economics, and potential audience size than do conventional (broadcast) means of reaching mass audiences.[52] Terrorists are already incorporating approaches to communication that provide many of these advantages, although in more limited form. For example, copying and distributing CDs (or, a few years ago, videotapes) is a crude form of podcast that is already a prevalent means of distributing recruiting materials, claiming responsibility for attacks, and disseminating propaganda messages. For these activities, advanced communication technologies could offer important benefits, although these benefits would seem to depend on the ability of advanced communication technology to reach greater audiences at lower cost or effort, rather than their potential for enhanced security.

Another possible change that may result from advances in network technologies may be a shift in the balance of emphasis between exploiting communication and exploiting stored

---

[51] In fact, the very nature of terror use by a few to influence the many implies that terrorists, by design, have chosen a mode of conflict that inherently makes very limited demands on communication for attacks.

[52] It is also true that reaching mass audiences is necessary to mobilize mass demonstrations and rallies. However, organizations that incorporate such operations in their strategies become much more difficult to characterize clearly as terrorist organizations. When an organization's positive appeal to a populace overshadows its need to use terror to influence the populace, the situation is perhaps better characterized as an insurgency. This situation differs sufficiently from operations by a terror group that it would not be appropriate to apply this analysis, despite any similarities between the two situations.

data.[53] To the extent that ubiquitous, strong encryption, dynamic mode switching, and similar tools increase the difficulty of taking advantage of communication, more emphasis may be placed on capturing and exploiting large caches of stored information, such as hard drives. From what is known through unclassified sources, intelligence services have exploited computers or hard drives used by Ramsey Yousef, Ksheik Mohamid, Abu Musab Zarqawi, and others after the devices fell into security forces' hands as a consequence of arrests, seizures, and, in one case, good journalism.[54] Investigators can also exploit the information stored on cellular telephones or other digital devices found on suspects, penetrating a group's network and degrading or even dismantling its operational capabilities. Although this might seem an obvious breach of security practice that a careful group would avoid, some groups have made such mistakes. For example, during the shootout with and subsequent arrest of a Red Brigades member in 2003, Italian police confiscated a Psion personal digital device, which had an address database, names, and other information useful in the subsequent arrests in October of that year (Ceresa, 2005, p. 206). However, media attention to the information and its exploitation by security officials has led to a rapid change in the security practices among the *Brigatisti*, the individual members of the Red Brigades. As cell phones become increasingly similar to computers in their capabilities, the opportunity and value of exploiting the information in such personal communication devices may grow substantially.

The last of the possible changes in communication practices and their effects on terrorist activities that we will discuss is enabled largely by the advent of a single network technology: autonomous networking.[55] As we discussed earlier, because autonomous networking can bypass centralized monitoring by security forces at the switch or through the intermediate organization that manages the network infrastructure, it can avoid the monitoring that security forces can conduct by controlling the switch or gaining the cooperation of the organization that manages the network. The primary effect of this capability may be to decrease the time it takes a terrorist group to set up a network of sufficiently secure communication and the resulting changes in terrorist operating style. If such technology has a popular, and thus sufficiently large, installed base, it may be easy for a terrorist group to set up temporary communication networks with no antecedents or traces of the network left afterward with third parties. This may be useful to help avoid surveillance by allowing meetings at a distance, validating parties prior to clandestine face-to-face meetings, providing added security for support activities that

---

[53] A similar shift in emphasis may occur in the balance between the intelligence assessment of the information within messages (which requires the ability to decrypt encrypted messages) and the assessment of information from message traffic profiles.

[54] See Cullison (2004) for insights into what can be available from such sources. The article describes how two journalists for *The Wall Street Journal*, Alan Cullison and Andrew Higgins, obtained and analyzed computer hard drives used primarily by Ayman al-Zawahiri. The stored information included nearly 1,000 text documents dating back to 1997, including budgets, training manuals for recruits, and scouting reports for international attacks. The documents shed light on everything from personnel matters and petty bureaucratic sniping to theological discussions and debates about the merits of suicide operations. The information also included video files, photographs, scanned documents, and Web pages, which illuminate a sophisticated effort to conduct a global, Internet-based information campaign.

[55] The kind of ubiquitous encryption that the research team believes is very likely in the future would improve the security of terrorist operations relying on this technology to the extent that encryption technology could be viewed as a necessity.

appear legitimate to casual observers such as purchasing equipment, or even for creating a temporary network to support just one particular operation. Absent some preparation to deal with this challenge, these communications will be largely undetectable to the security force.

But again, this is a technology that can prove useful to *both* terrorist organizations and security forces, depending on how well the users can understand and monitor the devices they are using. Devices that are sophisticated enough to embed autonomous networking capabilities pose the possibility that they are susceptible to, or even designed with, a covert capability that can be employed against terrorist groups. In the resulting measure-countermeasure contest, such technology guarantees neither side an inherent operational advantage, and both may end up with roughly the same balance with which they started.

Overall Effects of Changes in Communication Technology

In summary, terrorists are likely to use advanced communication technology in the future only when it suits their operational needs and when they judge the risks to be acceptable, rather than allowing technological capabilities to dictate operations. Good operational tradecraft is likely to determine and limit terrorist operations despite future technology-based communication capabilities, and it seems unlikely that even ubiquitous secure communication would alter the operational balance between security forces and terrorists in a profound way.

However, the changes in the network technologies used in communication are likely to be substantial, and they probably will have some important effects on other aspects of terrorist activities. Recruiting and propaganda activities may benefit greatly from advanced network technologies because of the capability that these technologies offer for selecting and communicating more directly with the specific audiences to whom terrorists wish to get their message, and security forces may increasingly find value in exploiting personal communication devices that store large amounts of data. Additionally, some new forms of autonomous, or self-forming, networks may offer substantial, perhaps even revolutionary, advantages, because they eliminate the need for users to depend on a network provider and thus they may eliminate the vulnerabilities that can arise from a terrorist group having to depend on those outside the terrorists' own organization.

Attack Operations

This discussion focuses specifically on the attack phase of operations and the role of network technology in enabling a group to attack a specific target or multiple targets either simultaneously or in a coordinated sequence.

As terrorist organizations are often under threat from security forces, the need for secrecy has dominated their ability to coordinate activities and launch operations, often limiting the complexity and the scope of the operations that an organization can carry out. In the past, the application of network technologies in operations has been limited, frequently due to security concerns. Transport and staging have largely depended on the terrorist organizations' ability to navigate to the proper position at a predetermined time and to communicate only when necessary. Although these limitations do provide a substantial degree of security for the terrorists' operation, they do constrain their ability to adapt to changing circumstances or to the reaction of security forces.

Current State-of-the-Art Operations

Several network technologies, most notably cell phones, remote garage door openers, and other communication devices, have enabled terrorists to improve their use of improvised explosive devices (IEDs). There have been well-publicized cases of terrorist use of cellular phones in operations (e.g., the March 2004 Madrid train bombings) (Raman, 2005; Dunnigan, 2005). Terrorists also use a number of other technologies for the purpose of detonating IEDs. Fuzing systems have been developed from pressure-activated devices (physical, water, or barometric), photosensitive cells, motion detectors, heat detectors, radiation detectors, electronic timers, and fuse wire. In preparing for his planned attack on Los Angeles International Airport, Ahmed Ressam purchased electronic components in September 1999 and assembled them as timing devices for his bombs (*United States v. Mokhtar Haouari*, S4 00 Cr. 15 [JFK] S.D.N.Y., July 3, 2001, p. 575). Remote detonation devices have also allowed terrorists to use some individuals as suicide bombers without their knowledge by using couriers to carry explosives that were then detonated remotely ("Conventional Terrorist Weapons," undated). They have also allowed the use of detonation schemes that are tailored to a target's signature or other operational parameters such as daylight or darkness.

Although a wide range of groups have used these technologies successfully, customizing specific electronic components for operational use may be difficult for a terrorist group unless it has a safe haven in which to operate. Although we often apply a "mirror-image" model to terrorists in our thinking, viewing them as animated by the same considerations, motives, and attractions as ourselves, terrorists are often less attracted to technological solutions than are well-funded commercial or military organizations. The environment in which they operate is, in some ways, much more constrained than that of security forces and very different from the environment in which much of our security establishment develops its equipment and conducts its training. Modifying an existing device or platform to meet a terrorist organization's operational requirements requires technical skill and resources, and the costs of failure can be much greater than the loss of the operatives who carry out an attack. Mistakes can lead to the compromise of the entire organization, so a small group with limited resources that finds itself confronting a numerically superior and well-funded security force is often very conservative—or, from its perspective, very practical—in its decisions.

Some groups have tried to modify electronic devices for use as detonators and failed;[56] others have had quick, easy, and successful experiences.[57] A terrorist group's success in developing effective weapons often depends on the group's willingness to devote the resources and time needed to do so—for example, by creating special organizations to carry out such activi-

---

[56] Author interview with law enforcement technical expert, Northern Ireland, May 2005.

[57] The Madrid bombings killed 191 people, the third-largest death toll from Islamic terrorism since September 11, 2001. The Madrid attack was put together in eight weeks, using stolen explosives and cell phone detonators assembled by one of the conspirators. It required no central direction from al Qaeda and no special technical expertise other than the skills that each of the locally recruited conspirators brought to the organization. See Windrem (2005).

ties. Particular devices may also vary in their applicability to a particular terrorist group's operations. Cell phones, for example, can be problematic given some unpredictability in the network delays that occur when a call is placed. Although a short delay before detonation may not affect many terrorist operations, they could be important for particularly time-sensitive actions such as hitting a moving target.[58]

Terrorist organizations have reportedly sought to adapt and use network technologies, including GPS, for remote guidance of weapon-laden vehicles. PIRA, for instance, reportedly experimented with retrofitting vehicles that could be guided by remote control (Harnden, 2000, p. 208) or by GPS navigation "somewhat like a pilotless cruise missile" (Geraghty, 2000, p. 212). The GPS technology has been cited as an example of convergence, and perhaps actual collaboration, between PIRA and other organizations such as Basque Homeland and Liberty (Euskadi Ta Askatasuna, or ETA) and the Revolutionary Armed Forces of Colombia (Fuerzas Armadas Revolucionarias de Colombia, or FARC), which also have reportedly experimented with these technologies.[59] Clear evidence is not available of more mundane uses of these navigation technologies for basic movement and logistics operations, but it is likely that terrorist organizations use them.

The Future of Terrorist Operations

Increasingly, as more functionality is integrated into personal electronic devices, clusters of technologies are likely to play an increasingly important role in enabling terrorist operations. Although these devices are referred to individually as smart phones, digital cameras, personal digital music devices, personal memory devices (e.g., jump drives), GPS devices, and digital video recorders, they increasingly embody clusters of technological capabilities that can be useful to terrorists.

The development trajectory of cell phones in particular may enable terrorists to have a single electronic device that can provide integrated, redundant mechanisms for signaling operations or detonating explosive devices.[60] These may include, for example, an electronic timer, a cell-phone arming or trigger device, and an integrated photo diode. These capabilities may also enable heretofore-unused triggering mechanisms, such as devices designed to tap into location-based services to enable a location-specific detonation capability that is not currently available in an integrated package.[61]

---

[58] Author interview with former military explosives expert, England, May 2005.

[59] Author interview with a former security forces member, England, March 2004.

[60] In contrast, for example, to the terrorist building in multiple, *separate* fuzing mechanisms (see, for example, Baker (2005).

[61] To date, there has been little evidence of terrorists using GPS location technology in weapons, and, in an era in which the use of suicide operatives increasingly characterizes terrorism, the advantages of GPS use might be limited. Signal availability may be one reason for the current lack of interest (e.g., limited signal strength in subways or inside steel-girder buildings); however, with the growing number of location-based services, this may no longer prove as restrictive. Triggering with a GPS-based device may also be both desirable and practical for some types of operations that pose problems for suicide bombers, such as the delivery of a radiological dispersal device (RDD) or a true nuclear device concealed in a shipping container that is subject to a very long transit time.

Furthermore, changing the functionality of many of these devices might require modifying only the software to achieve the desired function, which would be particularly attractive because modifying hardware that involves tightly integrated functions may be difficult.

Emerging technologies may also enable potentially new triggering mechanisms based on the use of RFID tags ("Benetton to Tag 15 Million Items," 2003) and readers (Balkovich, Bikson, and Bitko, 2005). An improvised device could be rigged with a fuzing apparatus designed to activate in the presence of a specific tag (e.g., unique tags implanted by a clothing manufacturer for inventory or after-sales tracking or a tag clandestinely planted on a vehicle or person.) Increasingly, wireless devices that uniquely specify an individual (e.g., Bluetooth-aware phones or WiFi PDAs) could also be used as triggering mechanisms for attacks on specific people, just as they now are used to target the theft of a celebrity's private phone list. In the future, the ability to read the unique information contained on these devices may enable triggers that are tailored to a specific individual because the information is uniquely associated with that person (e.g., the electronic ID number of the individual's phone or the contents of his or her address book).

As communication technologies become increasingly able to use available wireless communication modes, it may become more difficult to interrupt the triggering of IEDs. Early generations of terrorist detonation devices could frequently be jammed because they could only receive transmitted signals on a limited number of frequencies.[62] In the future, a single PDA or smart phone will likely be able to receive information using conventional cell-phone signals or WiFi, as well as Bluetooth or other available communication signals. The diversity of signals could make jamming devices based on these technologies difficult. Furthermore, in certain areas, wireless communication on those frequencies may be so critical to providing emergency services or other public or private functions that it may not be possible to jam them at all.

Though advances in network technology can provide operational advantages to terrorists, our research suggests that the effects of these changes will be more incremental than revolutionary. Continued improvements in smart devices and network technologies may enable increasing miniaturization with more secure and robust operation. However, although these changes might bolster terrorists' ability to stage attacks and sustain their operations over longer periods, the overall balance between the terrorists and security forces is not likely to be seriously affected. The existing means of delivering weapons and conducting attacks is adequately secure, practical, and effective, and major improvements to the existing capability do not seem to be strongly related to network technologies. Depending on the development trajectory for these technologies, terrorists may also need substantial technical expertise to successfully adapt consumer electronic and other network technologies to fulfill their needs and still ensure security. This requirement could also constrain these devices' potential value to some terrorist organizations.

---

[62] See, for example, discussion of such countermeasures in the conflict against the PIRA in Jackson (2005).

Propaganda and Persuasion

Acts of terror generally take place on a "stage," with one or more audiences in mind (Jenkins, 1975). Unlike most guerrilla attacks or special operations, the act of terror usually has little inherent military value, but instead sends a message to the target audience, for example, to draw attention to a historical grievance, demonstrate power, or to discredit authorities. This element of terrorism may be called propaganda or mass persuasion—influencing key audiences including the public and changing their attitudes, opinions, and behaviors—and is central to terrorist operations. Modern tools used for propaganda can include cell-phone camera images of violent acts, amateurish video clips, professionally produced videos, and even the simple fact that an iconic target was selected for attack, as well as more canonical propagandistic materials, such as the statement of selected religious authorities, claims of responsibility, and pronouncements of terrorist leaders.

Until recently, for propaganda and persuasion activities, terrorist groups were, in one important aspect, highly dependent on third parties: They did not usually own TV or radio stations (and only occasionally printing presses) and so required media organizations to spread their messages more broadly (one might, cynically, characterize the relationship as symbiotic, since news media are rarely reluctant to cover sensational news) (Wilkinson, 1997). These materials were occasionally distributed through national or international media (often after a sensational attack occurred) or by hand among supportive or tolerant audiences. Historically, the major channels for mass persuasion of which terrorist groups took advantage included television, newspapers, radio, and graffiti.[63] Although frequently serving their purposes, this relationship allowed terrorists little control over the content and framing of the messages that were communicated.[64]

Terrorist groups have internally reproduced and distributed manifestos, pamphlets, and video and audiotapes espousing their ideology and activities, for recruitment purposes or for communicating threats to adversary audiences. Al Qaeda operatives had planned to video-record the bombing of the USS *Cole* in 2000, but they were unable to get to their vantage point in time, *The 9/11 Commission Report* noted (National Commission on Terrorist Attacks upon the United States, 2004). Instead, the act was recreated and supplemented with training scenes. The report goes on to point out,

> Al Qaeda's image was important to Bin Laden, and the video was widely disseminated. . . . Al Qaeda members considered the video an effective tool in their struggle for preeminence among other Islamist and jihadist movements. (National Commission on Terrorist Attacks upon the United States, 2004, p. 191)

Until the past few decades, however, the parochial causes that terrorist groups supported, their poor in-house production capabilities, and their limited psychological knowledge meant

---

[63] The importance of graffiti should not be underestimated—in contrast to mass media outlets, it is a medium in which the terrorist does have control. A notable example of the effectiveness of the medium can be found in the notoriety and infamy of the murals in Belfast produced both by PIRA and its Loyalist terrorist opponents. See, for example, Jarman (1998).

[64] An excellent discussion of this topic in general can be found in Schmid and de Graaf (1982).

that, in general, their skill at propaganda and persuasion was usually of only modest effectiveness—either in generating support among friendly constituencies or in catalyzing political change in adversaries. PLO's and Hizballah's discovery that hijacking airliners and staging spectacular attacks could generate intense international interest may be seen as a rough starting point to the age of more sophisticated propaganda techniques.

Current State-of-the-Art Propaganda and Persuasion

Although some aspects relating to terrorist propaganda have remained largely the same—for example, the news media's appetite for sensational news—many network technology innovations and their commercial or societal responses, significantly change terrorist groups' ability to undertake information operations that attempt mass persuasion.

First, news coverage today is no longer in the control of just a few organizations (see Project for Excellence in Journalism, 2006). Many local and international outlets compete for viewers, and thus there is a nearly inexhaustible supply of willing partners to distribute terrorist messages. Further, many of these organizations offer access to distinct market segments: people who may speak particular languages, subscribe to particular religious beliefs, or have particular interests. This market differentiation greatly empowers any group that may get its message into the channel, because the group can target audiences of specific interest. For example, today it is possible for a terrorist group to issue a manifesto in one language to a given news outlet, but issue a different message, or a message with a different tone, in a different language to an altogether different news outlet.[65]

Second, terrorist groups are no longer dependent on a few media outlets: The Internet, in particular, allows terrorist groups or their supporters to have dedicated Web sites (such as al Battar for al Qaeda) and to post messages, videos, and the like for direct consumption by the public, without any need to employ independent media organizations, which might mistranslate or misstate the message.[66] Hizballah has been reported to own traditional media outlets including three radio stations, one television station, and two publications (Ranstrop, 1994), although it should probably not be considered typical of most terrorist organizations because of Iranian backing and its base in southern Lebanon.

Third, the sheer variety of media outlets has increased. Whereas, in the past, terrorists were limited to television, radio, graffiti, and the like, now they may also exploit the Internet for Web sites, e-zines, podcasts, and other emergent communication outlets.[67] Weimann (2004) has pointed out, "Al Qaeda combines multimedia propaganda and advanced communication technologies to create a very sophisticated form of psychological warfare." Chechen separatists have used videos to show that their fighters continue to carry out sabotage and other attacks to counter messages generated by Russia's state-controlled media.[68]

---

[65] Analyses of terrorist Web sites have reported differing content and tone between the English and Arabic versions of jihadi Web sites (Dartmouth College, 2003).

[66] The SITE (Search for International Terrorist Entities) Institute, a Washington, D.C.–based nongovernmental organization, monitors extremist Web sites and communiqués over the Internet on a daily basis.

[67] See, for example, Zanini (1999).

[68] Dartmouth College (2003). Video listings can be found at Kavkaz Center (undated).

Fourth, it is now easy to produce competent, even high-quality, materials for distribution. Inexpensive software and hardware allow documents, movies, music, role-playing games, and virtually every other form of communication to be crafted on a desktop computer or laptop. Moreover, current and emerging applications greatly reduce the difficulty associated with translation. For some types of communication, a multimedia document produced on a desktop may be translated with sufficient fidelity of meaning into any one of scores of languages[69] and immediately disseminated (Weimann, 2004). Multimedia itself, particularly that involving videos and pictures, need no translation.

Further, as discussed earlier, a wide variety of software tools can be used to forge or alter documents and images with relative ease.[70] Thus, terrorist groups may issue propaganda based on falsified scenes, recordings, and documents with only modest difficulty, and this production requires no special facilities or expert forgers and can be applied to live video broadcasts in near real time (with a delay of a fraction of a second) (see Amato, 2002). These capabilities can be combined to quickly field a sophisticated and successful terrorist information campaign. In the year and a half between the first videos attributed to Abu Musab Zarqawi's insurgent group in Iraq and Zarqawi's death, he used the Internet to combined written pronouncements, horrific videos, full-length propaganda movies, interviews with notable figures, and the drama of his negotiations and then swearing of allegiance to bin Laden into an information campaign that attracted worldwide attention.

Finally, it has become sharply more difficult to physically constrain terrorist propaganda. Although, in the past, governments have halted presses and shuttered theaters, choking off Internet-based or mass media communication is vastly more challenging. For example, with many news outlets available, it is difficult for authorities to prevent the dissemination of terrorist propaganda, particularly when the different news media have incentives to cover terrorist groups and their attacks. For example, despite German authorities' efforts to suppress the propaganda of the German group "2nd June," the news media often explained the derivation of the group's name in covering their activities; this, not incidentally, communicated the origins of their ideology and grievance to the public over and over again.[71]

One consequence of this has already resulted in a major shift in the conflict with terrorist groups. Since governments are far less able to restrict the dissemination of terrorist propaganda and can often no longer control the message of media outlets, they must instead combat propaganda with their own information campaigns. This is a new form of contest at which government leaders and agencies are not yet well practiced. It is inherently a challenging task—as

---

[69] See, for example, AltaVista (undated).

[70] For example, on December 12, 2000, at the request of the FTC, a U.S. District Court shut down the Web site of Jeremy Martinez of Tarzana, California, doing business as Info World. The FTC alleged that the defendant's Web site sold 45 days of access to fake ID templates for $29.99. The site contained templates for the creation of fake California, Georgia, Florida, Maine, Nevada, New Hampshire, New Jersey, Utah, Wisconsin, and New York driver's licenses. It also contained a birth certificate template, programs to generate bar codes—required in some states to authenticate driver's licenses—and a program to falsify Social Security numbers.

[71] For a primary source's discussion of this phenomenon, see Baumann (1979).

political campaign advisers have long known—the mere fact of addressing the terrorist's propaganda is likely to enhance its visibility and perhaps its influence.

The Future of Propaganda and Persuasion

We envision two worrisome possibilities for future terrorist propaganda trends: One is likely; the other less so.

First, as noted in the section on forging identities, we are on the brink of an era in which literally any video image may be falsified. Although techniques are available to verify the fraudulent nature of images, they are neither apparent nor likely to be available for many audiences. Given this capability, a technically savvy group could manufacture realistic video images of the President of the United States (or any other important international figure) speaking any words it wishes to put in his or her mouth and, given the trends in global information flow, could transmit these images far and wide very quickly. The possibilities are ominous: A public health official could be made to speak of a (nonexistent) biological attack, a diplomat made to derogate a friendly state or major religion, and so on. Even if quickly refuted, the effects may linger[72] and could undermine public confidence in subsequent official pronouncements. Such images may also be used to extend the terrorist group's mass appeal; for example, whether Osama bin Laden lives or dies, he may virtually issue proclamations for many decades.

The second development is perhaps less likely but worth considering: Terrorist groups may jam or even hijack U.S. information operations efforts. For example, if U.S. forces broadcast images of peace talks underway between warring religious factions, the technically savvy terrorist could rapidly create realistic (although fictitious) simulacra that show peace talks breaking down. Or if U.S. forces publish a glossy magazine and disseminate it, the terrorist group could quickly publish a nearly indistinguishable twin that violates local taboos or otherwise alienates the very population the United States is seeking to influence. Although there is nothing at all new about the aforementioned terrorist tactic (the name for this method, coined by social psychologist Robert Cialdini, is *poison parasite*), what would be new would be the speed and quality with which the copy could be produced and distributed to a wide audience.

In considering truly revolutionary developments in the realm of terrorist propaganda, few things could be more worrisome than terrorists acquiring an ability to hijack major media outlets. By *hijack*, we mean seize complete control of what the public was watching, irrespective of whether this was accomplished by physical, electronic, or other means. Depending on developments in network technologies both applied by the terrorist and used by media outlets in producing the programming, advances in network technology could enable such a scenario.

If a terrorist group could introduce its own imagery, documents, and narratives into the hijacked outlet, the results could be dramatic. For instance, should terrorists seize control of an evening news program, even if only for minutes, and manage to create the illusion that the mayor of New York City was announcing that the city had suffered a massive chemical attack,

---

[72] See Cialdini (1993); Petty and Cacioppo (1996); and Zimbardo and Leippe (1991) for overviews of this large body of scientific literature.

it is very likely that substantial disruption would ensue.[73] Even if order were quickly restored, doubts about the veracity of subsequent broadcasts might linger, and the confidence in the political leaders involved in handling the matter might be damaged. Used in conjunction with a physical attack that capitalized on whatever confusion resulted from the media hijacking, the fraudulent information might significantly impede the authorities' ability to respond to the physical attack—particularly in light of government agencies' growing dependence on the news media for information and for communicating with the public in such situations. Even without the use of a coincident physical attack, such media hijacking might be used as a weapon to produce lasting effects. For example, a sophisticated campaign of repeated hijackings could conceivably damage public confidence in some nations' political leaders to the point that people might come to believe that they needed to rely on themselves or militia-like organizations for protection. As with most aspects of the conflict with terrorist groups, the outcome would depend largely on how well the authorities could respond to repeated incidents of this sort, but, if their performance were inadequate, the effects might be so great as to destabilize a city or locality, thus furthering the terrorist group's aims.

Which of These Network Technologies Are Potentially Most Attractive to Terrorists?

Of the many possible network technologies that could be used in terrorist activities, which will be most attractive to potential adversaries? We base our assessment on the expectation that a terrorist will adopt a technology if it can confer one of two types of benefits with reasonable risks. These benefits are

1. those that improve the organization's ability to carry out activities such as recruiting and training that are relevant to its long-term ability to survive and conduct operations
2. those that improve the outcome of its immediate attack operations.

In this model of decisionmaking, a group adapts to the operational situation it faces so it will survive and be successful in its mission. For technology choices, each group will define and evaluate the benefits and risks associated with mission success and group survival in a manner that its culture and leadership determine. We can only approximate such decisions here by assuming that terrorist groups make rational decisions over the long run using these group sur-

---

[73] Orson Welles' *War of the Worlds* radio broadcast, which took place on Halloween night of October 1938, provides an example of what might happen with the insightful use of network technology. Although three out of four families in the United States owned a set, many people were not yet fully attuned to how the medium could be used. Welles (at the time a young, but insightful, broadcaster without great standing) tapped into the fears of a nation on the eve of World War II and convinced thousands that Martians were invading the United States. The public reaction can be described as disruption, and perhaps even panic, despite the prior publication of the story behind the broadcast (Wells, 1898). For detailed information and documentation about the broadcast, see Gosling (undated).

vival and mission success criteria.[74] In some situations, the benefit of novelty may outweigh the risks associated with a new technology.[75] In others, a group's contacts with other groups that have been successful in using a particular network technology may influence the decision.[76] In still other situations, existing and familiar technologies may be sufficient despite the availability of new equipment, if they offer sufficient versatility and flexibility. Further, different groups will likely adopt different technologies reflecting their individual circumstances and perceived needs. As a result of such factors, it is not possible to *predict* which network technologies terrorists will adopt in the future; however, we can make some educated assessments about which network technologies appear to offer what could be interesting combinations of benefit and risks given the insights we have about future technology and about terrorist activities.

Network Technologies That Can Enhance Strategic or Enabling Activities
Of the current and emerging network technologies discussed in the previous chapter, several appear potentially attractive to terrorist groups as a result of their effect on the groups' strategic or support activities. These technologies include

- virtual gaming technology for recruitment
- cyberpayment systems that can be used anonymously by any bearer for fund transfer
- massively multiplayer games for training and populace influence
- ubiquitous replication of high-quality forged credentials
- impersonation of key persons in electronically mediated individual communication
- worldwide, secure, multimode mobile data and voice communication
- falsified video and audio avatars of leadership figures for public propaganda
- electronic hijacking of news media outlets.

With these technologies, the terrorist may gain a useful set of tools, previously available only to larger and more established organizations. These tools could allow terrorist organizations to work faster and use fewer people without the costs of either recruiting and training larger numbers of operatives and supporters in country or moving people in and out of the United States. The use of these tools would also decrease the burdens associated with some

---

[74] For a discussion of benefits and risks from the terrorists' perspective and how its culture and leadership may influence this, see Jackson (2001). A terrorist organization's core culture and its operational style help determine whether or not an organization is attracted to new technologies, but they are seldom an overt part of the leadership's deliberate decisionmaking process. Although each situation is unique and each organization's leadership influences how culture and style are considered with respect to technology, if the new capability is perceived as incompatible with the organization's view of itself, it is unlikely that the changes necessary for technology adoption will be made. For example, Ceresa (2005, p. 220) describes the Italian terrorist group the Red Brigades (BR-PCC, Brigate Rosse per la Costituzione del Partito Comunista Combattente) as holding a strong image of its role and mission, which is defined by its past exploits, which in this case includes kidnappings and murders. She reports that the group determined that cyberattacks, with their indirect effects on society and state institutions, were unattractive to the group, in contrast to direct physical action.

[75] Advanced weapons like antiaircraft missiles provide a useful example: Clearly their value to some groups is sufficient to produce some use, though the spotty results groups have gained by doing so emphasizes some of the risks of new technologies. See, for example, Schaffer (1998).

[76] See Cragin et al. (2007) for a more extensive discussion of relying on outside groups for technology acquisition.

types of training and allow specialized knowledge workers to operate from more secure locations while helping field teams operating closer to the targets and in attack operations. However, these tools are likely to have only an indirect effect on attack outcomes and so the direct risk they pose to homeland security might be characterized as low.

Network Technologies That Can Enhance the Direct Outcomes of Attacks

The main risk from terrorist organizations is still primarily their use of direct, violent, physical attacks against personnel and infrastructure.[77] These "kinetic" attacks can produce effects with very high confidence, are difficult to defend against, and are very efficient in destroying targets and creating the desired effects of a terror campaign. The attack, along with the information activities that amplify or augment it, is the primary tool terrorists use to influence people and events.

The technologies we have assessed offer few reliable improvements for attack operations themselves. Why? Simply put, the ability to coordinate operations in the dynamic manner that network technologies enable do not matter a great deal in deliberate, well-planned, and well-rehearsed operations conducted by modestly well-trained personnel against static, often defenseless civil targets.

Dynamic responsiveness matters a great deal to security forces that must react to an attack, and highly skilled military forces can use network technologies to great effect in fluid battles. But this is not the situation that terrorists face, and we should not rely on a model of terrorist operations that envisions them as soldiers without uniforms. In the conflict in which we are engaged, the terrorists are not in a circumstance in which their operational gains from the use of network technologies will be significant relative to what they already can accomplish using far simpler approaches that capitalize on their inherent advantages of choosing the time, place, and manner of attack to maximize their effectiveness.

There are some important exceptions to this conclusion; most notably, uses of network technology such as remote command detonation and sensor-initiated detonation of explosive devices do offer improved attack capabilities to terrorists in a number of circumstances.

---

[77] Other efforts under this research study examined potential advances in weapon technology and found them to have compelling evidence that they could directly pose substantial new risks in the future. See Bonomo et al. (forthcoming).

CHAPTER THREE

Security Force Responses to Terrorists' Acquisition and Use of Network Technologies

In the previous chapter, we examined the ways in which network technology can enable terrorist operations, what future network technologies might offer to terrorist organizations that adopt them, and which technologies appear particularly attractive for terrorist activities.[1] Now, we will turn to terrorist use of these technologies from security forces' perspective and adversaries' potential responses to their use.

Determining how to respond to the use of specific network technologies is more difficult than assessing which technologies might be attractive to terrorist groups. In addressing this issue, the research team decided that a countermeasure's contribution to the homeland security or counterterrorist mission's success was the most useful measure for determining the value or priority of a countervailing action.

Prioritizing based on a countermeasure's value (its contribution to the success of the homeland security or counterterrorist mission) may not result in the same choice of countermeasures as a scheme of priorities based on a technology's value to the terrorists. In particular, the approach that considers the contribution to the security force mission takes into account how effective a countermeasure may be in addition to how valuable or attractive a targeted technology is to terrorists. For example, the greatest benefit to the security force mission may not lie in simply disrupting the terrorists' use of network technologies; it may lie in exploiting the information that these technologies store, work with, and communicate to develop actions that more directly affect terrorist organizations, such as attacks on a cell's base or the arrest of its members.

In this chapter, we describe a method for making decisions about countermeasures and other defensive responses to terrorist adoption of these technologies on the basis of their practicality, viability, and payoffs. Making these judgments requires examining

1. *the role a specific network technology plays within a terrorist group's overall technology strategy*—i.e., is the technology one among many options available to the group or a con-

---

[1] We would like to stress the important distinction among the activities for which terrorists typically use network technologies. These are the activities that support attack operations—not the attack operations themselves. In the analysis presented in the previous chapter, we found that the use of network technologies usually allows fewer terrorists to do what once required a larger organization and thus improves the group's overall efficiency. As a result, network technologies improve a group's efficiency and therefore its survivability, but they typically do not directly improve attack outcomes to any significant degree.

stant element in many things the group does? Payoffs for implementing countermeasures would be much larger in the latter than the former situation.

2. *the balance of benefits and risks of technology use from both the terrorists' and security forces' perspectives.* As a measure of the value of countering technology use to the mission accomplishment, we will use an estimate of the benefits and risks similar to the judgments that decisionmakers might make to evaluate the mission-related payoffs that result from alternative countermeasure options when used against the different terrorist technology strategies.

3. *the countermeasure options available to security organizations.* The potential payoff of interfering with a terrorist organization's use of a specific technology is irrelevant unless countermeasure options exist to actually do so. The last part of our analysis examined the options available to security organizations for doing so and assessing their practicality against classes of network technologies.

When combined, these components can define a framework that will allow us to compare the relative payoff for each combination of network technology strategy and countermeasure.

The Role of Specific Network Technologies Within Terrorist Groups' Technology Strategies

Although there is broad agreement that terrorist organizations adapt and evolve over time, there is a good deal of uncertainty about how these groups choose to adopt and use a technology because the forces that shape those decisions are not easy to observe. In explaining the technology choices that terrorists might make, some analysts focus on novel or advanced weapons and suggest that these characteristics are a sufficient motive for choosing one technology over another.[2] Others point out that most terrorists appear to be operationally conservative, relying on basic technologies such as guns and explosives, which implies that novelty alone may not be a sufficient reason to choose a particular technology.[3]

Previous RAND research has examined the role of technology in terrorist groups, the basic actions that a group must carry out to adopt a new technology (Jackson, 2001), and processes of organizational learning and change in terrorist enterprises.[4] These studies have shown that specific technologies play different roles within the activities and operations of individual terrorist organizations. For example, some groups make extensive use of single technologies—either because they have built up significant expertise in them or because they represent versatile tools that can be used in many different ways—while others draw on many different technologies to support their organizational and offensive activities.

---

[2] See, for example, Jenkins (1975, p. 15); Bell (1987, p. 50).

[3] For example, see the discussion in Hoffman (2001).

[4] In particular, Jackson, Baker, et al. (2005a, 2005b), include several case studies specifically contrasting the learning over time of some well-studied terrorist groups.

We use the concept of organizational *technology strategies* to define a simple framework to summarize the approaches that terrorist groups might take in adopting network technologies. The resulting framework summarizes four broad approaches that terrorist groups take with respect to new technologies:

1. *Specialize in specific technologies, enabling the group to customize and shape them to the needs of its activities and operations.* Typically, implementing such an approach requires specialization by some parts of the organization for the acquisition and employment of such technology.

2. *Adopt many technologies, providing the group with a wide variety of options to apply as needed.* Although variety-based strategies do not necessarily require groups to build up specialization or deep knowledge of particular technologies, groups must invest time and resources in maintaining their ability to use many different technologies well. Variety-based strategies are made much easier when technologies are readily available on the commercial market.

3. *Focus on individual technologies, but choose technologies that are versatile and can be used in many different ways.* The more ways in which an individual technology can be used, the higher its potential value to an individual terrorist group. The ubiquity of communication across the terrorist activity chain—and the availability of these technologies on the commercial market—demonstrates that many network technologies could constitute very versatile technologies within these groups' operations.

4. *Rely on technology opportunistically, without a concerted organizational focus on adopting and deploying novel technologies.* Just because technologies appear potentially attractive to terrorists, there is no certainty that they will adopt them. Although passing up opportunities to use new technologies will deny organizations their benefits, such a strategy may also result in little organizationwide vulnerability to technology failures, countermeasures, or exploitation.[5]

The significant differences in the potential place for specific network technologies within a terrorist group's technology strategies will frame the level of payoff for countermeasure efforts targeting those technologies.

Specialization in individual technologies is necessary for terrorist organizations to respond to some types of change in their environment. For example, as a result of jamming of radio detonators for their bombs, terrorist organizations have been forced to make electronic modifications to circumvent the countermeasures. Doing so requires a level of specialized understanding about the details of remote detonation systems; a group without that knowledge could not adapt. However, specialization requires resource commitments to develop and maintain requisite expertise and knowledge. Committing resources requires confidence that the effort and cost is the only way to produce acceptable results. In such circumstances, the group's image of itself is often convolved with its decision to acquire powerful technology, as was the case for Aum Shinrikyo, the Japanese terrorist group that developed nerve gas for an attack on

---

[5] This is discussed in more detail in the comparison of costs, risks, and benefits of these technologies for terrorist groups.

the Tokyo subway in 1995. The investments that groups must make to specialize require that groups will do so primarily when they see a significant payoff for use of the technologies and, therefore, suggest that the value of countering the technology use will be higher for security organizations. Although few of the network technologies we examined appeared to have revolutionary effects on terrorist operations or effectiveness, those with that potential would require some specialization by the group (e.g., creation or subversion of massively multiplayer games for their own purposes or real-time falsification of video feeds).

If a specific network technology is only one element within a technology strategy built on a *variety* of different technologies, then the effects of countermeasures may be quite different. Variety-based strategies focus on groups acquiring many different technologies that can then be applied as needed to situations in which they are appropriate. Given the number of network technologies that the consumer market already provides, the barriers to groups pursuing variety-based strategies are significantly lowered. The consumer electronics and computer market increasingly makes network technologies, such as mobile computing, digital image manipulation, and wireless communication, affordable and easy to obtain. Terrorists already use such off-the-shelf network technologies for functions such as communication, planning, or producing video-based training materials. The market also provides a considerable variety of infrastructure services—which rely on facilities and equipment that might be potentially exploited, including use of the Internet or interrupting or even hijacking news media broadcasts. In such situations, where a group only has to make a limited investment to acquire a technology and has many alternatives that can perform similar functions, the value of efforts to counter the use of specific technologies will be both transient and limited.

When individual technologies can be used in many different ways, groups can rely on their *versatility* to provide them with significant flexibility and utility. Terrorist preferences for guns and bombs—two versatile weapon technologies—are a good example of this behavior. Such robust, but rudimentary, technologies provide versatility because they can be applied to a broad set of uses or situations and reliably deliver sufficient results most of the time.[6] The wide use that terrorists make of video recording technology for tactics development, training, and propaganda provides an example of the versatility-based approach. The robust effects of such devices, coupled with the acceptable levels of risk typically associated with their use, means that these technologies are sufficient for many tasks. For example, nearly any handheld video recorder has adequate capability for most of the main terrorist activities for which these devices are used, such as reconnaissance, training, tactics development, and propaganda. Equipment (or weapons) based on such technology can often be used even if circumstances change. This flexibility enables rapid adaptation to changes in the operational environment, though generally at some lesser level of capability than more specialized systems provide. Such versatile technologies are often found in the equipment produced for commercial use and are available through the consumer market.

---

[6] We are using *rudimentary* here in a relative sense. Typically, no network technology is thought of as rudimentary, but, in relative terms, it can be. For example, for video recording, nearly any video recorder, including now-obsolete-to-the-consumer VHS machines, would have adequate utility for the purposes for which terrorists use them and offer a simplicity of operation that allows nearly anyone to operate them without complicated training.

Our analysis found that the advantages brought about by these types and uses of network technology do allow fewer terrorists to perform certain activities and so they improve the group's overall efficiency, but they are not likely to have great impact on the outcome of terrorist attacks because these activities are involved with supporting, rather than conducting, attack operations. In short, they may improve a group's efficiency and therefore its survivability, but not its operational effectiveness to any significant degree. From the perspective of considering countermeasures, even if versatile technologies do not themselves make major impacts on the effectiveness of individual terrorist operations, their use in many different ways and contexts can mean that their cumulative benefits over time can be very large. This suggests that countermeasures aimed at denying use of these technologies—if such measures are available—could be valuable because of their broad-based effects on a range of group activities.

Though our report has focused on the effects of groups pursuing and acquiring new technologies, it is important to remember that not all groups will necessarily choose to pursue even seemingly valuable technologies. Even for network technologies that appear attractive to terrorists, some groups may not choose to pursue them or use them only *opportunistically* when they can do so at limited cost and effort; the technologies they do use are likely to have a much more limited effect on activities and operations. Such an approach generally avoids potential vulnerabilities resulting from the technology, which could be a target for security force countermeasures or exploitation. From the perspective of defensive planning, this situation defines a lower bound at which only limited investments should be focused on countering technology use.

By highlighting four distinct technology strategies and corresponding roles for individual network technologies within those strategies, we do not mean to imply that terrorist organizations must choose a single approach to acquiring and using new technologies. Organizations can pursue more than one simultaneously, e.g., training much of the group in versatile technologies while selected elements specialize in particular technologies such as command detonation devices. We should also stress that these strategies pertain to a group's overall approach to technology and there will be exceptions in specific situations. As a result, because terrorists are keenly aware that using some technologies may result in vulnerabilities that can compromise their security. There may be technologies that they will never acquire regardless of the technology strategy they use (discussed in more detail in the next section).[7]

Benefits and Risks from Network Technology Use

The second ingredient in assessing how to most effectively counter terrorist adoption of network technologies is the net balance of benefits and risks from their use by terrorists—from both the adversary's and the defense's point of view. Table 3.1 details a simple framework that addresses the benefits and risks for both the terrorist organization and security forces arising from the use

---

[7] This would likely be the case even when the group makes only opportunistic use of technology; such an approach would not imply that the group takes a casual approach to decisions involving technology and their operations, especially if there are security issues involved.

Table 3.1
Risks and Benefits of Network Technologies to Terrorist Organizations and Security Forces

Network Technology Categories	Terrorist Organizations		Security Forces	
	Benefits	**Risks**	**Benefits**	**Risks**
Connectivity technologies (Wireless communication modes)	Efficiency gains, operational flexibility. Can provide anonymity and security under some circumstances	Interception and signal interference at a distance, difficult to verify correctness of system, dependent on external providers	Interception opportunities, tainted equipment	Enemy can act faster, operate over wider area, fewer personnel involved
Personal electronic devices (e.g., cell phones, PDAs)	Efficiency gains from having information available, real-time adaptation, removing ambiguity (passing images rather than descriptions)	Interception of communication internals and externals, unintended or induced emissions for fingerprinting	New target for exploitation, increases tendency to use communication devices	Allows enemy to have more information available
Software and applications	Can increase autonomy for terrorist organizations—generation of new products. Allows for greater operational performance. Facilitates cyberattack and cybercrime	Diverts resources from primary mission—e.g., potentially alters the tooth-to-tail ratio. Can lead to undesired dependency on technology. Introduces possibility of applications revealing sensitive information or locations	Opportunity for introducing compromised software. Provide potential signature of terrorists	Enables terrorist self-help. Prevents security focus on critical suppliers of information. Enemy might have very different appraisal of situation from defender's. Introduces greater variability into adversary planning
Information technology services and access to the Internet	Facilitates borderless operations	Facilitates bidirectional access (computer network attack or exploitation)	Can allow for borderless counterterrrism operations	Adversary operations exploit jurisdictional and bureaucratic boundaries
Mobile computing	Facilitates more survivable distributed operations	Can lead to loss of critical data and to operational compromise	Chance encounters can provide wealth of data. Mobile protection lower than for safe-houses	Harder to roll up organization with a single geographically focused operation
Video and other recording devices	Intelligence and training opportunities	Less thinking and distillation of tactics and more glitz. OPSEC compromise	Can provide insights into enemy training and operations	Enemy can disseminate information faster, and with greater efficacy

of network technology and summarizes the outcome of our examination of current and future network technologies. Relying on this framework, we can assess the broad themes of payoff that the use of network technologies entails for terrorists and for security forces.

Benefits and Risks of Using Network Technology for Terrorist Groups

Our analysis of future uses of network technology did not find compelling evidence that the use of these technologies would be remarkable enough that they might be used solely for their novel effects. Thus, from the perspective of terrorist groups, the benefit from network technologies is associated with three measures: conducting an operation with a higher likelihood of success, enabling a mission that heretofore was highly impractical, or increasing efficiency to the point at which more operations can be executed with the limited trained personnel pool available.

A quick assessment of the terrorist organizations' benefits in Table 3.1 reveals that efficiency gains and facilitating long-distance, undisclosed-location, or more autonomous operations account for the majority of the benefits that arise from the use of these technologies.

A quick review of the terrorist organizations' risks shows that the downside in virtually all cases is associated with the risks to a terrorist group's operational security. Some risks are associated not with the technology itself, but with the processes and procedures that must be adopted to use it effectively. Standardized concepts and the consequent routines, rules, and categories enable data storage, access, and communication across interdependent but otherwise separated individuals. Successfully implementing and using such standard practices requires resources and time for training and may also define patterns that law enforcement and intelligence organizations may exploit. In addition, risks may also include the difficulty of finding people capable of using the technology correctly and the danger of depending on an external producer. It is worth noting that, in addition to being problems in and of themselves, these two difficulties can also contribute to failures in operational security for the terrorist organization.

This concern about the organization's security underscores one key weakness of terrorist organizations that may be important when considering security force countermeasures. Although a particular terrorist operative or even an entire group may conduct an operation in a high-risk manner, terrorist organizations must be highly sensitive to risk at the level of the network itself and at key nodes—such as a leadership group—if they are to survive to pursue their causes. This vigilance is needed because, even though terrorist organizations can be resilient, they ultimately have much more limited resources than security forces do, and compromising the network or key elements of the network could rapidly undermine organizational capabilities.[8]

Benefits and Risks to Security Forces of Terrorist Use of Network Technology

The calculus for the security force is, of necessity, more complex because the terrorist organization enjoys a last-move advantage and can also choose the mission area for which the technology is used. However, when terrorist groups do pursue new technologies, not all the consequences will be negative for security organizations: New technology systems may provide new opportunities to track or identify terrorist activities. To evaluate countermeasure strategies, the security forces must consider not only the risks and benefits that the terrorists' use of network

---

[8] Hoffman notes that 50 to 75 members of the terrorist group the Red Brigade imposed a multiyear reign of terror on Italy; and for more than 30 years, a dedicated cadre of approximately 200 to 400 IRA gunmen and extremists tied up a major portion of the British Army in Northern Ireland (Hoffman, 2004, pp. 13–14).

technologies imposed on them, but also the likely risks and benefits under which the terrorists will operate should countermeasures be put into effect.

An examination of the security forces' benefits in Table 3.1 reveals that all but one of the benefits to security forces deal with enhanced opportunities to exploit either the technology being used by the terrorists or the information that is stored, processed, or transmitted by the network technology that the terrorists are using. From the point of view of the security forces, all but two of the risks deal with enhanced efficiency on the terrorists' part. The remaining two deal with another supporting function; they offer strengthened security to the terrorists. None of the risks directly relates to the attack phase of operations.

Some finer points that arise when examining the table carefully should be specifically pointed out. First, some of the network technologies examined may have a balance of benefit and risk (positive from the terrorist organization's perspective and negative from that of the security force) to potentially warrant a countermeasure program designed to preclude, degrade, or eliminate the technology or its use. Certainly, one such example of a technology that provides more than modest benefits to terrorists is remote detonation of explosive devices. Second, overt exploitation of network technologies to enhance security forces' direct action against the terrorists (such as offensive operations or arrest and prosecution) have a dual effect: Exploitation not only enables the direct action, but it also will likely decrease the exploited technologies' attractiveness to the terrorist organizations, once exploitation is suspected. This diminution in attractiveness may produce a deterrence effect that drives terrorists away from using the technology in question.

Options for Countering Terrorist Use of Network Technologies

Available insights on how individual network technologies fit into terrorist groups' technology strategies and the balance of the risks and benefits of their acquisition can provide a basis for identifying situations in which attempting to hinder terrorist use could be valuable. However, moving from theory to practice requires understanding what options exist to do so, their feasibility, and practicality. We have structured our examination of this dimension of the security planning problem around four major classes of security action: attempting to deny the terrorist access to or use of a network technology, trying to counter operational use of the technologies, seeking to exploit their use of it, or—when better options are not available—not attempting to counter the technology and focusing resources on other priorities where the payoff would be higher.

Options Relying on Denial. Denying terrorists the use of network technologies by interdicting their acquisition of technology or through direct countermeasures to the specific technical equipment might appear to be an attractive option. Doing so would seem to prevent terrorists from using the technology in any way to strengthen their organizations—for example, by recruiting new members or providing training—as well as preventing it from being used to help carry out attacks.

Interdicting the acquisition of a technology by terrorist groups is particularly problematic in the case of network technology because terrorist organizations often base their acquisition strategies for these technologies either on the variety provided by the consumer market or on versatile technologies that perform well enough in most circumstances, including in the presence of countermeasures. In all but the most restrictive security settings, it appears to be unrealistic to rely on keeping network technologies, particularly those available as consumer or commercial equipment, out of terrorist hands in an attempt to interdict technology acquisition.

Options Relying on Countering Operational Use. Technical countermeasures, such as jamming communication signals or attacking terrorists' Web sites with malicious software, are eminently feasible against commercial- or consumer-based network technologies and their associated hardware, but it is not clear that the benefit is worth the risk (which can include the disruption of legitimate users of the technology). The core of the problem lies again in implementation and the fact that the conflict with terrorist groups is fought in open society, not on an isolated battlefield. As a consequence, the technical countermeasures that are most interesting are those that can be employed with a high degree of specificity that allows them to be targeted in a way that can reduce the unintended consequences. For example, in the case of cell phones, extremely short-range and highly directional disruption could be advantageous if the target can be identified, but anything else would so disrupt normal cell phone use that it would probably be impractical.

In such a conflict, the real problem is identifying the target for the countermeasure, not the technical capability to deny the use of network technology. This identification is difficult, and such operations often require sophisticated targeting or detection resources that are currently available only to a few organizations, particularly if the employment concept requires security forces to wait until the terrorist group begins communication before countermeasures can be initiated.

Such uses of direct technical countermeasures tend to be episodic and largely determined by whether or not the terrorists expose themselves, their communication, or their Internet activities as targets. This means that some portion of terrorist communication can continue without disruption with the result that there may be insufficient hindrance to a terrorist group's ability to plan, coordinate, or conduct operations. Such employment is also problematic because there are many means available to terrorist organizations that allow them to perform adequately even if countermeasures are successful. For example, shutting down a Web site used by terrorists is technically well within current security forces' capabilities, but re-instituting the Web site is equally within terrorist capabilities, resulting in a move-countermove cycle with little consequence over time.

In short, countering terrorists' use of network technologies through technical countermeasures that directly target the equipment or its technology is likely to be effective only in specific circumstances, such as supporting or leveraging other operations. As a result, this countermeasure is likely to be only moderately effective overall.

Options Based on Exploitation. The fact that denying terrorists use of network technologies appears infeasible in most circumstances is not an argument for doing nothing in response to their acquisition and deployment. The modest, though real, improvements in efficiency these technologies provide does suggest that groups will continue to pursue them. An option that might prove more advantageous than trying to keep network technologies out of terrorists' hands or direct technical countermeasures would be to exploit the terrorist use of network technologies.

If the primary goal of security forces is to defeat a terrorist organization in the long run, there may be opportunities to turn the terrorist organization's use of network technology tools to the security force's advantage by exploiting the information that network technology collects, stores, and transmits to enable attacks, arrests, and other direct actions against the terrorist group.

By using network technologies to evaluate targets, communicate, remotely detonate devices, and other activities, terrorists play in many ways to the strengths of security forces, who are generally resourced so they have more and better equipment and so they have access to more technically skilled personnel. Further, no matter how robust the terrorist organization is as a whole, each cell is extremely fragile. The more a cell extends by networking, the further away it gets from a system in which trusted internal security officers keep its members, equipment, software, communication, and other assets and actions under close observation. As the distance between the center of a terrorist cell, where planning and policymaking take place, and its operational fringe increases, the likelihood that terrorist activities or the terrorist organization as a whole will be compromised by human intervention or the subversion of its technology also increases.[9]

Security forces might, for instance, subvert the terrorists' technology by introducing compromised cell phones or computers, which would permit the covert collection of information used by or in close proximity to the device.[10] The nontransparent functioning of many technology devices enhances such an operation's feasibility; without significant technical training and resources, users cannot really understand how these technologies operate, which reduces the likelihood that the user could discern that security forces are exploiting the technology. Prudent terrorist organizations would be very concerned about such possibilities because they can directly threaten a central operational imperative—maintaining the organization's security. To protect themselves, they may be disinclined to take full advantage of what network technologies have to offer, minimizing their use of network technology tools outside protected environments where the system's physical security can be assured and where remote compromise of the system is difficult. It is also possible that they could find it necessary to choose an operational profile that is more vulnerable than they would be able to use if the technology were viewed as trustworthy.

This suggests that that there may be a useful distinction between the network technologies that terrorists would choose due to their versatility and those for which they would rely on the variety provided by the market. Both are problematic if not difficult to counter in acquisi-

---

[9] A network is no more secure than its least secure element, and ensuring that the parties accessing the network are valid users is a significant security concern for network administrators. Even sophisticated commercial computer security systems may not guard against the physical duress of parties accessing the network or against hardware compromise as a means of extracting information. As a consequence, terrorist groups that rely on the consumer electronics and computer markets may find such consumer-acceptable limitations to be a fatal flaw for the environment in which they operate.

[10] More complicated variations on this theme are possible (although they may be more difficult to employ), e.g., covertly encouraging terrorists to adopt compromised consumer systems or particular technologies that are more easily exploited by security forces. To keep the discussion focused on the main ideas, we will mention such variations only in passing.

tion or during employment. Network technologies valued (and acquired) for their versatility, may be well understood to their users, depending on how rudimentary they actually are. But technologies whose strength is market-based variety are often more opaque to users, and this opens up the possibility for a sophisticated security force to take advantage of aspects of their software or hardware of which users may not be aware.

Options That Focus Resources on Other Problems. In addition to these options for security forces, there is a final option—choosing to focus not on any aspect of the use of network technology, but rather on other weaknesses of terrorist organizations that promise better payoffs. This option would be a logical choice if the impact of terrorist use of an information technology were limited or hard to attribute to network technologies in any useful way.

Evaluating the Countermeasure Options

The preceding discussion outlines a framework for evaluation with two dimensions: (1) the types of technology strategies that a terrorist group might use to acquire network technologies coupled with the comparative benefits and risks of each strategy and (2) the countermeasure options available to security forces. This gives us a framework in which to compare the payoff for each combination of effect and countermeasure.

This framework can be visualized by relating the different terrorist technology strategies to the different strategy options for countering the technology's use in a matrix. This is illustrated in Table 3.2, which summarizes the key aspects of the analysis to this point. The table arrays rows of the different terrorist technology strategies and their broad effects on security (*major* for specialized strategies in which terrorist have built significant skill in a given technology, *moderate* for versatility and variety strategies in which individual technologies either have more modest effects or are but one among many options available to the group, and *minor* for opportunistic strategies in which groups may only use a technology episodically), against columns of the different countermeasure options available to the security forces. These are arrayed against the countermeasure options available to security forces: interdicting technology acquisition, countering the operational use of the technology, exploiting terrorists' use of the technology, and focusing on other priorities.

Table 3.2 also specifies the payoff we found in our analysis for each combination of terrorist technology strategy and the security force countermeasure options.[11] Although the framework in the table may be used to assess any form of technology, we apply it to network technologies. The analysis relating to network technologies is discussed in the sections that follow.

---

[11] The table presents the payoff to security forces based on a general consideration of the technical feasibility and costs associated with countermeasure options for a broad category of network technologies (e.g., those consistent with a specialization strategy because of their complexity and potentially high payoff to terrorists). With a more detailed assessment of these factors, the preferred action for a specific given technology, as differentiated from a general technology strategy, might be different.

Table 3.2
Payoffs to Security Forces of Counters to Network Technologies

Terrorist Technology Strategy	Effect on Security if Technology Use Is Successful	Security Forces' Countermeasure Options			
		Keep Technology Out of Terrorists' Hands	Develop Measures to Counter Operational Use of Technology	Allow Terrorists to Use Technology and Exploit	Focus on Other Counterterrorism Priorities
Specialization	Major change	Preferred	May be impractical	Not acceptable	—
Versatility (rely on robust but rudimentary technology)	Moderate change	May be impractical	Moderate payoff	Limited payoff	—
Variety (rely on consumer market)	Moderate change	Impractical	Limited payoff	Preferred	—
Opportunism	Minor change	—	—	—	Preferred

Network Technologies Within Specialized Technology Strategies

Technologies whose use by terrorists would result in major changes in security would almost certainly require security forces to choose a counter intended to keep the relevant technologies out of terrorist hands and to develop measures to counter their use in the field.

As reported earlier, the research team found no strong argument that a future use of network technology would result in a revolutionary change in the relationship between security forces and terrorist groups to produce a major effect on homeland security. However, some of the uses of network technology that were assessed to be particularly valuable to terrorist organizations do have some potential, however small, to cause such major changes, and it is instructive to consider them in the framework we have set up. These include

- virtual gaming technology for recruitment
- cyberpayment systems that any bearer can use anonymously for funds transfer
- massively multiplayer games for training and populace influence
- ubiquitous replication of high-quality forged credentials
- impersonation of key persons in electronically mediated communication
- worldwide, secure, multimode mobile data and voice communication
- falsified video and audio avatars of leadership figures for propaganda
- electronic hijacking of news media outlets.

These applications would most likely require most terrorist groups to specialize to achieve the skills necessary to be successful. In particular, the possible effects of terrorists' use of cyberpayment systems, ubiquitous high-quality forgery, impersonation of key decisionmakers in electronic communication, and posing as news media outlets would seem to indicate that the acquisition interdiction strategy would be the preferred strategy should these uses of network technology prove more likely than our analysis estimates them to be, and it would be prudent to carefully monitor the development of such network technologies.

As Table 3.2 illustrates, our analysis suggests that, unless there is stronger information indicating that revolutionary uses of network technologies are closer to reality, investments of time and resources greater than those required to keep a wary eye out for such developments do not seem warranted.

Network Technologies Within Versatility- and Variety-Based Strategies

The situation differs markedly for the situations in which terrorists rely on versatile technologies or the variety of network technologies that the commercial market continuously offers. Typically, these are associated with moderate changes in security. Many network technologies fall into this strategy category.

Our analysis finds that, in these situations, network technology is unlikely to significantly change operational outcomes, but most practical uses of network technology by terrorists produce real improvements in a group's efficiency, allowing smaller organizations to conduct more effective terror campaigns.

In the preceding discussion, we noted that it is useful to distinguish between two cases based on whether they were based on very versatile technologies or a great variety of technologies. Although, in either case, pervasive market forces and many vendors pose problems for acquisition denial, there are some differences in their susceptibility to operational countermeasures, and the situation for exploitation is markedly different.[12]

Technology Strategy Based on Versatility. The "always adequate, but seldom spectacular" characteristic of technologies acquired for a strategy that relies on a technology's versatility implies that the payoff to security forces from exploiting such robust but rudimentary technologies is likely to be limited in comparison to the damage that these technologies reliably produce. This is due to the uncomplicated way in which these technologies produce their operational effects and how familiar users can be with the equipment based on simple technologies (and how familiar they can be with how it might be exploited). It is also due to the fact that modifying key aspects of older-generation technology once a device is in terrorist hands usually requires physical access to the device, unlike newer-generation networked devices, which may be able to be modified remotely.

Our analysis suggests operational countermeasures (including technical countermeasures designed to defeat the equipment itself) are the most attractive option for countering these technologies. But this is a choice among poor alternatives. The very nature of these rudimentary technologies implies that they will work "well enough" and "most of the time." As a result, they do not lend themselves well to either countermeasures or exploitation.

Technology Strategy Based on Variety. For network technologies that are associated with an acquisition strategy that relies on the variety of the consumer market, Table 3.2 illustrates that our analysis found that the denial strategies (interdicting technology acquisition by terrorist groups and using technical countermeasures to thwart the use of terrorists' equipment) promise only modest payoffs—if they are even possible. Interdicting the consumer market

---

[12] As discussed previously, this is related to the role that the market plays in terrorist acquisition of technology and, thus, indirectly, to how quickly the market changes the technology in users' hands and, as a consequence, whether or not terrorist users can be completely aware of the ways in which security forces can exploit such technology.

acquisition of such technology by these groups is probably impractical. Because of the role that these technologies play in the terrorist activity chain (they are typically used for supporting activities), technical countermeasures to the use of equipment based on these technologies is likely to have limited payoff in terms of lowering the risk, vulnerability, consequences, or threat of terrorist attack.

However, Table 3.2 also illustrates that our analysis found that exploitation countermeasures have a number of attractive aspects for this situation. Because these network technologies typically deal with information and this may be the key to compromising a group, the payoff to security forces for exploiting the terrorists' use of network technologies may greatly outweigh the risk of allowing terrorists to use them. Because of the fast pace of change brought about by the market and the fact that much of the equipment now offered to consumers is highly complex and is often remotely accessible, terrorist users may not be aware of how sophisticated security forces can exploit the equipment or its use.

Additionally, such exploitation would threaten one of the fundamental operational imperatives of terrorist organizations—their security. The common risk to a terrorist group in each of the categories of network technology that we examined was the possibility that using the network technology might compromise the group. Threats to such fundamentals of terrorist operations as security demand a response that would involve personnel, material resources, and leadership attention, thus detracting from other operational activities. The possibility that security forces can exploit network technologies may in itself discourage terrorist groups from using them because the potential risk to terrorist groups of the technology being exploited may be too high.

Network Technologies Pursued Opportunistically

Although the information presented in the framework regarding this strategy is simple, the point to be made is important for determining effective ways to counter terrorists' use of network technologies. Most of the countermeasure options do not apply to this strategy (see Table 3.2); however, the point is that, if the effect of the use of network technology is minor, then the best choice from a security force perspective is to focus resources on other aspects of the conflict.

Although this appears to be a point that is easily understood because of the stark way that it is presented in the framework, in actual practice, other factors may obscure the situation. These include the sometimes-distorting emphasis that news media coverage can give to terrorist use of a technology even if it is sporadic and there is no evidence that it is particularly effective, the importance accorded such use by those uncomfortable with technology, and the emphasis that some opinion makers may place on preventing terrorist groups from using particular forms of technology regardless of the efficacy of such an approach.

Countermeasure Approach Suggested by the Evaluation

The analysis suggests that the approach to countering terrorist groups' use of network technology should focus primarily on the use of the technology as an efficiency-enhancing mechanism rather than one that allows dramatic new operational effects. Those efforts that do focus on the "revolutionary" potential of network technologies might usefully be oriented on the detection and warning of network technology uses with such potential, rather than investing in countermeasures to the technology itself.

In developing such a strategy, security force decisionmakers should consider not only *denial countermeasures*—that is, measures that preclude the technology's adoption, prevent its use, or degrade an adversary's ability to use it as intended—but also *exploitive countermeasures* that enable security force operations that disrupt a terrorist organization more directly through offensive operations or arrests.

Our evaluation of the countermeasure options also echoes a point made earlier in reporting on the risk and benefit analysis conducted for the different types of network technology: Security forces would do well to consider a countermeasure strategy based on terrorist organizations' preference for exploiting the use of network technologies, rather than seeking to counter them directly.[13]

---

[13] Because technology-based assessments often focus on technical questions and focus little analytic attention on how terrorists might use new technologies in their operations, the countermeasure response that such assessments often imply is simply to directly counter the technology in question. From a technical perspective, the approach suggested here, which can include allowing terrorists to use a given technology in order to exploit it, may seem counterintuitive, but may be the most effective (and practical) option in some circumstances.

CHAPTER FOUR

Conclusions and Recommendations

Conclusions

Our analysis of network technologies, the needs of terrorist organizations, the actions they pursue, and how they acquire network technology and carry out their operations leads us to the following broad conclusions.

Major Breakthroughs in Terrorist Attack Operations?

The small size of terrorist operations and the deliberate way in which terrorists plan those operations leads to a highly scripted execution process that does not require advanced network technologies beyond what is readily available to them now.

Versatility, Variety, Efficiency, and Effectiveness

Terrorists have shown that they can adapt easily to new technologies such as widespread Internet access. We expect that they will readily embrace network technology that is widely available and that enables them to carry out their activities with fewer people or lower risk of compromise. Key technologies that fall into this category include use of advanced Internet services, versatile use of video recording, production of false documents, targeted dissemination of information to audiences, use of cyberpayments for funds transfer, and use of smart cell phones or other personal devices that package clusters of network technologies and sensors.

Precluding Terrorists from Getting Technology and Developing Direct Counters

Technologies that feature versatility and variety are largely driven by the worldwide consumer market. It is not practical to keep these kinds of technologies out of the hands of terrorists. Such technologies as digital video recorders and smart cell phones can simply be bought off the shelf in the advanced economies that are typically the terrorists' targets, in states that support and sponsor terrorism, and even in the failed states that may serve as their bases of operations. Prudent planning and strategy must assume that, if terrorists want them, they can get them. Even if it were possible to deny terrorists these technologies, the benefits of doing so would be only marginal because of the many alternative ways to perform the terrorist support activities for which these technologies are used and the indirect effects that such support activities have on attack operations outcomes.

Exploitation Seems the More Promising Option

The best use of resources for those attempting to counter terrorist operations would seem to be developing ways to exploit the network technologies that terrorists will continue to use and that offer the highest payoff. As is the case with most people who use cell phones and computers, most terrorists do not have detailed knowledge of how those devices work or their security arrangements. Therefore, it may be possible to alter them in ways that enable security services to identify the users or their locations or to monitor their transmissions. Such exploitation can support direct action, such as arrests, and, because it threatens a key operational imperative of terrorist organizations, their security, it can also deter the use of the technology.

Security Services' Role

One area that might require careful monitoring would be network technologies that could enable terrorist organizations to take over or temporarily pose as established media outlets. Even though it is unlikely that they could do this for a sustained period, even a short takeover could be highly disruptive, particularly in densely populated urban areas. Other technological capabilities that should be monitored because of their potential effects include the use of cyber-payment systems, ubiquitous high-quality forgery, and the impersonation of key decisionmakers in electronic communication.

Recommendations

Terrorist groups' decisions to adopt a new technology and security forces' countervailing decision about how to counter the terrorists' choice may be usefully likened to a high-stakes two-sided game with advantage determined and redetermined over and over through a series of moves or initiatives. In the long term, the advantage in this interaction goes to the side that can adapt to the changes introduced by the other or by serendipitous events more quickly, thereby limiting whatever windows of advantage those changes open. This adaptation can be undertaken in an ad hoc manner relying on an organization's inherent characteristics, such as small size or short lines of communication. But it is more likely to be consistently successful over the long term, particularly for a large or complex organization, if the nature of the longer-term contest with terrorist groups is appreciated and a process for determining a series of appropriate responses (or preemptive actions) is designed and put into place. Our recommendations about what should be done to accomplish this follow.

Design a System to Address Terrorist Use of Network Technologies

Consistently choosing good approaches to counter terrorists' use of network technologies over time requires a system that can do the following:

- determine whether terrorists have adopted or are likely to adopt a new network technology
- identify the new technology's likely effect on the terrorist group's operational capabilities
- select a strategy and response

- marshal the resources needed to develop and field the approach to counter operational use of the technology
- do these in a timely enough manner that it can have an effect on an adaptive adversary.

Acquire and Retain People Who Can Make the System Work

Homeland security forces and other organizations involved in combating terrorism need the following core competencies to address terrorist organizations' use of network technologies:

- an understanding of the technologies themselves, particularly the technical challenges of exploitation and the operational limitations imposed by terrorist and security force operations
- the ability to track terrorist adoption, use, or avoidance of particular technologies
- the capability to determine which responses, or which mix of them, is most appropriate in light of security forces' goals
- the capability to develop plans and execute operations to carry out the selected responses as part of the larger strategy to counter terrorist organizations.

Take the Initial Steps Needed to Implement Such a System Promptly

Determining who should be responsible for these activities is complex. For example, DHS might be the best agency to execute some of these actions, but, for others, DHS might be more effective as an advocate or coordinator for actions that would best be accomplished using the core competencies of other agencies. Thus, at issue is both who should accomplish these activities *and* who should define and coordinate the activities; it is likely that it will take some time to determine these responsibilities.[1] However, carefully chosen initial activities by DHS and other homeland security entities can quickly provide a good basis for a system that can counter terrorist organizations' network technology use. These include DHS activities that

- *continue and accelerate the recruitment, retention, and professional education of technically skilled personnel who understand network technologies*
- *define the requirements for intelligence collection that is focused on terrorist use of network technologies and communicate them to the intelligence community.* Defining the needs for the kind of information needed to counter and exploit terrorist use of network technologies requires coordination with the Department of Defense, other law enforcement agencies, and private-sector entities due to their unique understanding of the problem, their own operations, and potential exploitation measures. Once the needs have been defined, they must be passed to the intelligence community as collection requirements.
- *include an examination of the terrorist use of network technologies in the homeland security research program.* This recommendation ties closely with the preceding one; it argues for a specific effort within homeland security research to focus on network technologies. Such

---

[1] For similar homeland security matters involving cross-agency capabilities (e.g., infrastructure protection), the President has made the appropriate assignments of responsibility and directed that mechanisms such as the Homeland Security Council be used for interagency coordination. See, for example, Bush (2003).

a program should explore both recent experience with terrorist use of network technologies and the potential use of such technologies in the future.

- *develop the capability to determine which of the following responses is warranted when terrorists use network technologies:*
  - *Exploit the use of the network technology*; for the technologies that fall into this category, coordinate with the other intelligence and law enforcement agencies to put in place the procedures necessary to incorporate such exploitation into ongoing intelligence, military, or law enforcement operations.
  - *Develop and employ operational countermeasures to the network technology;* for the technologies that fall into this category, institute the means through which the countermeasure is developed and law enforcement or other security forces are trained in its use.
  - *Disrupt the process by which terrorist groups acquire new network technologies*; for the technologies that fall into this category, institute the cross-agency mechanism necessary to undertake a disruption strategy, including associated intelligence requirements, commercial control regimens, and related day-to-day operational aspects of the homeland security mission such as customs enforcement or infrastructure protection and monitoring.
  - *Determine that investment and effort are more effective in other counterterrorism efforts.*

- *develop a capability to respond quickly with technical and engineering solutions to counter or exploit emerging network technology that terrorists are using.* Network technology developments occur relatively rapidly, and terrorist organizations have shown that they can adopt new technologies quickly. This results in a need for a rapid-response capability. Some consideration should be given to using the practical and time-sensitive engineering approach that NASA uses in addressing system problems during space flight missions as a model for such a rapid-response capability. Particular attention should be paid to diffusion of the new technologies among security partners and approaches to encourage state and local government entities to adopt newly developed responses in a timely manner.

Although fielding the system that can determine appropriate responses to terrorist organizations' use of network technologies in the manner that this analysis explains and illustrates is a complex, multiagency undertaking, the recommendations outlined herein should provide a basic capability within DHS that can contribute to the homeland security mission in the short term and is capable of being shaped to provide the most efficient and effective ways to address this threat over the longer term.

Bibliography

Adams, James, *The Financing of Terror: Behind the PLO, IRA, the Red Brigades, and M-19 Stand the Paymasters: How the Groups That Are Terrorizing the World Get the Money to Do It*, New York: Simon and Schuster, 1986.

"All Over Bar the Shouting? Islamist Terrorism," *The Economist*, August 6, 2005.

AltaVista, "Babel Fish Translation," undated Web page. As of January 15, 2007:
http://babelfish.altavista.com/

Amato, Ivan, "Lying with Pixels," *Technology Review*, January 11, 2002. As of February 20, 2007:
http://www.technologyreview.com/Infotech/12115

"Ambush in Afghanistan," video posted by aliveleak on January 17, 2007. As of February 20, 2007:
http://www.liveleak.com/view?i=0339b32a48

Argyris, Chris, and Donald A. Schön, *Organizational Learning: A Theory of Action Perspective*, Reading, Mass.: Addison-Wesley Co., 1978.

Armstrong, Ilena, "Hactivism: Protest or Petty Vandalism?" *SC Magazine*, September 2002.

Arquilla, John, and David Ronfeldt, eds., *Networks and Netwars: The Future of Terror, Crime, and Militancy*, Santa Monica, Calif.: RAND Corporation, MR-1382-OSD, 2001. As of January 15, 2007:
http://www.rand.org/pubs/monograph_reports/MR1382/

———, eds., *In Athena's Camp: Preparing for Conflict in the Information Age*, Santa Monica, Calif.: RAND Corporation, MR-880-OSD/RC, 1997. As of January 15, 2007:
http://www.rand.org/pubs/monograph_reports/MR880/

———, *The Advent of Netwar*, Santa Monica, Calif.: RAND Corporation, MR-789-OSD, 1996. As of January 15, 2007:
http://www.rand.org/pubs/monograph_reports/MR789/

Associated Press, "FBI E-Mail Monitor to Be Monitored," *CBS News*, August 10, 2000. As of August 3, 2005:
http://www.cbsnews.com/stories/2000/08/10/tech/main223613.shtml

———, "Alleged Suspects Shot 'Casing Video' of Capitol," wcbstv.com, April 28, 2006. As of January 15, 2007:
http://wcbstv.com/911/local_story_118222108.html

Australian Mobile Telecommunications Association, "More Mobiles Than Fixed Phones in China," November 25, 2003.

Baker, John C., "Jemaah Islamiyah," in Brian A. Jackson, John C. Baker, Peter Chalk, Kim Cragin, John V. Parachini, and Horacio R. Trujillo, *Aptitude for Destruction*, Vol. 2: *Case Studies of Learning in Five Terrorist Groups*, Santa Monica, Calif.: RAND Corporation, MG-332-NIJ, 2005, pp. 57–92. As of January 8, 2007:
http://www.rand.org/pubs/monographs/MG332/

Baker, John C., Beth E. Lachman, Dave R. Frelinger, Kevin M. O'Connell, Alexander C. Hou, Michael S. Tseng, David T. Orletsky, and Charles W. Yost, *Mapping the Risks: Assessing the Homeland Security Implications of Publicly Available Geospatial Information*, Santa Monica, Calif.: RAND Corporation, MG-142-NGA, 2004. As of January 8, 2007:
http://www.rand.org/pubs/monographs/MG142/

Balkovich, Edward, Tora K. Bikson, and Gordon Bitko, *9 to 5: Do You Know if Your Boss Knows Where You Are? Case Studies of Radio Frequency Identification Usage in the Workplace*, Santa Monica, Calif.: RAND Corporation, TR-197-RC, 2005. As of January 15, 2007:
http://www.rand.org/pubs/technical_reports/TR197/

Bassham, Lawrence E., and W. Timothy Polk, *Threat Assessment of Malicious Code and Human Threats*, National Institute of Standards and Technology Computer Security Division, March 10, 1994. As of January 15, 2007:
http://csrc.nist.gov/publications/nistir/threats/

Baumann, Michael, *Terror or Love? Bommi Baumann's Own Story of His Life as a West German Urban Guerrilla*, New York: Grove Press, 1979.

BBC News, "Mobiles Outstrip India Landlines," July 2, 2004. As of January 15, 2007:
http://newswww.bbc.net.uk/2/hi/business/3860185.stm

Bell, J. Bowyer, *The Gun in Politics: An Analysis of Irish Political Conflict, 1916–1986*, New Brunswick, N.J.: Transaction Books, 1987.

———, *The Secret Army: A History of the IRA*, Dublin, Ireland: Poolbeg, 1998.

"Benetton to Tag 15 Million Items," *RFID Journal*, March 12, 2003. As of January 15, 2007:
http://www.rfidjournal.com/article/articleview/344/1/1/definitions_off

Birchall, D. W., J. J. Chanaron, and K. Soderquist, "Managing Innovation in SME's: A Comparison of Companies in the UK, France and Portugal," *International Journal of Technology Management*, Vol. 12, No. 3, 1996, pp. 291–305.

Birkler, John, C. Richard Neu, and Glenn A. Kent, *Gaining New Military Capability: An Experiment in Concept Development*, Santa Monica, Calif.: RAND Corporation, MR-912-OSD, 1998. See Appendix C on radar resonance detection of weapons. As of January 15, 2007:
http://www.rand.org/pubs/monograph_reports/MR912/

B'nai B'rith, *Dangerous Convictions: An Introduction to Extremist Activities in Prison*, New York: ADL, 2002. As of January 8, 2007:
http://www.adl.org/learn/Ext_terr/Dangerous_Convictions.pdf

Bonomo, James, Giacomo Bergamo, Dave Frelinger, John Gordon IV, and Brian A. Jackson, *Stealing the Sword: Limiting Terrorist Use of Advanced Conventional Weapons*, Santa Monica, Calif.: RAND Corporation, forthcoming.

Brodie, Bernard, "Strategy as a Science," *World Politics*, Vol 1, No. 4, July 1949, pp. 467–488.

Burns, Tom, and G. M. Stalker, *The Management of Innovation*, London: Tavistock Publications, 1961.

Bush, George W., *December 17, 2003, Homeland Security Presidential Directive/HSPD-7: Critical Infrastructure Identification, Prioritization, and Protection*, Washington, D.C.: White House Office of the Press Secretary, 2003. As of February 20, 2007:
http://www.whitehouse.gov/news/releases/2003/12/20031217-5.html

"Business—Mobile Phones and Development: Calling an End to Poverty," *The Economist*, Vol. 376, No. 8434, 2005, pp. 51, 53.

Byman, Daniel, *Deadly Connections: States That Sponsor Terrorism*, Cambridge and New York: Cambridge University Press, 2005.

Campbell, Duncan, "How the Terror Trail Went Unseen," *Telepolis*, October 8, 2001. As of January 10, 2007:
http://www.heise.de/tp/r4/artikel/9/9751/1.html

Campen, Alan D., Douglas H. Dearth, and R. Thomas Goodden, eds., *Cyberwar: Security, Strategy, and Conflict in the Information Age*, Fairfax, Va.: AFCEA International Press, 1996.

Ceresa, Alessia, "The Impact of 'New Technology' on the 'Red Brigades' Italian Terrorist Organisation," *European Journal on Criminal Policy and Research*, Vol. 11, No. 2, 2005, pp. 193–222.

"Chechen Ambush," video posted by desantnik, December 31, 2006. As of February 20, 2007:
http://www.liveleak.com/view?i=a0b19b5640

Chen, Rongxin, "Technological Expansion: The Interaction Between Diversification Strategy and Organizational Capability," *The Journal of Management Studies*, Vol. 33, No. 5, 1996, pp. 649–666.

Chen, Yuxin, Chakravarthi Narasimhan, and Z. John Zhang, "Individual Marketing with Imperfect Targetability: Being Imperfect in Targeting Is Perfect for Profit," *Marketing Science*, Vol. 20, No. 1, Winter 2001, pp. 23–41.

Cialdini, Robert B., *Influence: Science and Practice*, 3rd ed., New York: HarperCollinsCollegePublishers, 1993.

Clutterbuck, R., "Trends in Terrorist Weaponry," in Paul Wilkinson, ed., *Technology and Terrorism*, London, UK, and Portland, Ore.: F. Cass, 1993, pp. 130–139.

———, *Terrorism in an Unstable World*, London and New York: Routledge, 1994.

Coll, Steve, and Susan B. Glasser, "Terrorists Turn to the Web as Base of Operations," *The Washington Post*, August 7, 2005, p. A1.

Collins, Eamon, and Mick McGovern, *Killing Rage*, London, UK: Granta Books, 1998.

"Conventional Terrorist Weapons," United Nations Office on Drugs and Crime, undated Web page. As of January 15, 2007:
http://www.unodc.org/unodc/terrorism_weapons_conventional.html

Cragin, Kim, "Hizballah, the Party of God," in Brian A. Jackson, John C. Baker, Peter Chalk, Kim Cragin, John V. Parachini, and Horacio R. Trujillo, *Aptitude for Destruction*, Vol. 2: *Case Studies of Organizational Learning in Five Terrorist Groups*, Santa Monica, Calif.: RAND Corporation, MG-332-NIJ, 2005, pp. 37–56.

Cragin, Kim, Peter Chalk, Sara A. Daly, and Brian A. Jackson, *Sharing the Dragon's Teeth: Terrorist Groups and the Exchange of New Technologies*, Santa Monica, Calif.: RAND Corporation, MG-485-DHS, 2007. As of July 23, 2007:
http://www.rand.org/pubs/monographs/MG485/

Cragin, Kim, and Sara A. Daly, *The Dynamic Terrorist Threat: An Assessment of Group Motivations and Capabilities in a Changing World*, Santa Monica, Calif.: RAND Corporation, MR-1782-AF, 2004. As of January 8, 2007:
http://www.rand.org/pubs/monograph_reports/MR1782/

Cragin, Kim, and Scott Gerwehr, *Dissuading Terror: Strategic Influence and the Struggle Against Terrorism*, Santa Monica, Calif.: RAND Corporation, MG-184-RC, 2005. As of January 8, 2007:
http://www.rand.org/pubs/monographs/MG184/

Cullison, Alan, "Inside Al-Qaeda's Hard Drive," *The Atlantic Monthly*, September 2004, pp. 55–72.

Daly, Sara A., and Scott Gerwehr, *Al-Qaida: Terrorist Selection and Recruitment*, Santa Monica, Calif.: RAND Corporation, RP-1214, 2006. As of February 20, 2007:
http://www.rand.org/pubs/reprints/RP1214/

Dam, Kenneth W., and Herbert Lin, *Cryptography's Role in Securing the Information Society*, Washington, D.C.: National Academy Press, 1996.

Dartmouth College, *Examining the Cyber Capabilities of Islamic Terrorist Groups*, Hanover, N.H.: Institute for Security Technology Studies at Dartmouth College, November 2003. As of January 8, 2007: https://www.ists.dartmouth.edu/TAG/ITB/ITB%5F032004.pdf

DARWARS, *Department of Defense Game Developers' Community*, undated Web page. As of January 8, 2007: http://www.dodgamecommunity.com/

Davis, Catherine, "Afghans Remember Slain Resistance Hero," *BBC News*, September 9, 2002. As of January 10, 2007: http://news.bbc.co.uk/2/hi/south_asia/2245456.stm

Denning, Dorothy Elizabeth Robling, *Information Warfare and Security*, New York: ACM Press, 1999.

————, *Activism, Hacktivism, and Cyberterrorism: The Internet as a Tool for Influencing Foreign Policy*, San Francisco, Calif.: Nautilus Institute, December 10, 2001. As of January 10, 2007: http://www.nautilus.org/archives/info-policy/workshop/papers/denning.html

Denning, Dorothy E., and William E. Baugh, Jr., "Cases Involving Encryption in Crime and Terrorism," The Cryptography Project, Georgetown University, Department of Computer Science, October 10, 1997. As of January 10, 2007: http://www.cosc.georgetown.edu/~denning/crypto/cases.html

Desmond, Michael, "Why Not WiMax," *Government Computer News*, Vol. 24, No. 11, May 16, 2005a. As of January 15, 2007: http://www.gcn.com/print/24_11/35773-1.html?topic=tech-reprt

————, "Broadband Bonanza," *PC World*, June 14, 2005b. As of January 15, 2007: http://www.pcworld.com/news/article/0,aid,121394,00.asp

Donahue, Arthur, "Terrorist Organizations and the Potential Use of Biological Weapons," in David W. Siegrist and Janice M. Graham, eds., *Countering Biological Terrorism in the US: An Understanding of Issues and Status*, Dobbs Ferry, N.Y.: Oceana Publications, 1999, pp. 21–35.

Dunnigan, James F., *The Next War Zone: Confronting the Global Threat of Cyberterrorism*, New York: Citadel Press, 2002.

————, "Iraqi Terrorists and the War on Terror," *Strategy Page*, March 3, 2005. As of January 15, 2007: http://www.strategypage.com/dls/articles2005/2005332.asp

Emerson, Steven, "Testimony of Steven Emerson Before the House Committee on Financial Services Subcommittee on Oversight and Investigations, 'Patriot Act Oversight: Investigating Patterns of Terrorist Fundraising': Fund-Raising Methods and Procedures for International Terrorist Organizations," February 12, 2002. As of January 8, 2007: http://www.au.af.mil/au/awc/awcgate/congress/021202se.pdf

Emery, Theo, "Video Research at MIT Puts Words into Mouths, with Startling Results," Associated Press, June 30, 2002.

Environmental Health Safety, "Air Modeling Software," Web page, November 27, 2006. As of January 10, 2007: http://www.ehsfreeware.com/amodclean.htm

Fair, C. Christine, "Militant Recruitment in Pakistan: Implications for al Qaeda and Other Organizations," *Studies in Conflict and Terrorism*, Vol. 27, No. 6, November–December 2004, pp. 489–504.

Farrell, Nick, "Mobile Phone Detects Bad Breath," *The Inquirer*, September 22, 2004. As of January 15, 2007: http://www.theinquirer.net/?article=18613

Federal Bureau of Investigation, "Indictment of Zacarias Moussaoui," press release, Washington, D.C., December 11, 2001. As of January 8, 2007: http://www.fbi.gov/pressrel/pressrel01/mueller121101.htm

"Federal Communications Commission Releases Data on High-Speed Services for Internet Access," press release, Federal Communications Commission, Washington, D.C., July 7, 2005. As of January 15, 2007: http://hraunfoss.fcc.gov/edocs_public/attachmatch/DOC-259870A1.pdf

"Five Days in an IRA Training Camp," *Iris*, Vol. 7, November 1983, pp. 39–45.

Freeh, Louis J., *Testimony of Louis J. Freeh, Director, FBI, Before the United States Senate, Committees on Appropriations, Armed Services, and Select Committee on Intelligence, May 10, 2001: Threat of Terrorism to the United States*, May 10, 2001. As of January 15, 2007: http://www.fbi.gov/congress/congress01/freeh051001.htm

Freeman, Christopher, and Luc Soete, *The Economics of Industrial Innovation*, Cambridge, Mass.: MIT Press, 1997.

Gade, Lisa, "Palm Treo 650 Palm OS Smartphone Review," *MobileTechReview*, May 11, 2005. As of January 15, 2007: http://www.mobiletechreview.com/Treo_650.htm

Gellman, Barton, "Cyber-Attacks by al Qaeda Feared: Terrorists at Threshold of Using Internet as Tool of Bloodshed, Experts Say," *The Washington Post*, June 27, 2002, p. A1.

Geraghty, Tony, *The Irish War: The Hidden Conflict Between the IRA and British Intelligence*, Baltimore, Md.: Johns Hopkins University Press, 2000.

Gerwehr, Scott, and Russell W. Glenn, *The Art of Darkness: Deception and Urban Operations*, Santa Monica, Calif.: RAND Corporation, MR-1132-A, 2000. As of January 8, 2007: http://www.rand.org/pubs/monograph_reports/MR1132/

———, *Unweaving the Web: Deception and Adaptation in Future Urban Operations*, Santa Monica, Calif.: RAND Corporation, MR-1495-A, 2003. As of January 8, 2007: http://www.rand.org/pubs/monograph_reports/MR1495/

Goffman, Erving, *Behavior in Public Places: Notes on the Social Organization of Gatherings*, New York: Free Press of Glencoe, 1963.

Goldberg, Ian, "Privacy-Enhancing Technologies for the Internet, II: Five Years Later," undated. As of January 10, 2007: http://www.cypherpunks.ca/~iang/pubs/pet2.pdf

Goldberg, Ian, David Wagner, and Eric Brewer, "Privacy-Enhancing Technologies for the Internet," January 21, 1997. As of January 10, 2007: http://www.cs.berkeley.edu/~daw/papers/privacy-compcon97-www/privacy-html.html

Gosling, John, "War of the Worlds: Invasion: The Historical Perspective," undated Web page. As of January 15, 2007: http://www.war-ofthe-worlds.co.uk/war_worlds_orson_welles_mercury.htm

Gunaratna, Rohan, *Inside al Qaeda: Global Network of Terror*, New York: Columbia University Press, 2002.

Harnden, Toby, *"Bandit Country": The IRA and South Armagh*, London: Coronet Books, 2000.

———, "Video Games Attract Young to Hizbollah," *The Daily Telegraph* (London), February 21, 2004, p. 18.

Heaton-Armstrong, Anthony, Eric Shepherd, and David Wolchover, eds., *Analysing Witness Testimony: A Guide for Legal Practitioners and Other Professionals*, London: Blackstone Press, 1999.

Hinnen, Todd M., "The Cyber-Front in the War on Terrorism: Curbing Terrorist Use of the Internet," *The Columbia Science and Technology Law Review*, Vol. 5, No. 5, April 19, 2004, pp. 1–42. As of January 8, 2007: http://www.stlr.org/html/volume5/hinnen.pdf

Hoffman, Bruce, *Terrorist Targeting: Tactics, Trends, and Potentialities*, Santa Monica, Calif.: RAND Corporation, P-7801, 1992. As of January 8, 2007:
http://www.rand.org/pubs/papers/P7801/

———, *Inside Terrorism*, New York: Columbia University Press, 1998.

———, *Terrorism and Weapons of Mass Destruction: An Analysis of Trends and Motivations*, Santa Monica, Calif.: RAND Corporation, P-8039-1, 1999a. As of January 8, 2007:
http://www.rand.org/pubs/papers/P8039-1/

———, "Terrorism Trends and Prospects," in Ian O. Lesser, Bruce Hoffman, John Arquilla, David Ronfeldt, Michele Zanini, and Brian Michael Jenkins, *Countering the New Terrorism*, Santa Monica, Calif.: RAND Corporation, MR-989-AF, 1999b, pp. 7–38. As of January 8, 2007:
http://www.rand.org/pubs/monograph_reports/MR989/

———, "Change and Continuity in Terrorism," *Studies in Conflict and Terrorism*, Vol. 24, No. 5, 2001, pp. 417–428.

———, *Insurgency and Counterinsurgency in Iraq*, Santa Monica, Calif.: RAND Corporation, OP-127-IPC/CMEPP, 2004. As of January 8, 2007:
http://www.rand.org/pubs/occasional_papers/OP127/

Horchem, H. J., "West Germany's Red Army Anarchists," in William Frank Gutteridge, ed., *Contemporary Terrorism*, New York: Facts on File, 1986, pp. 119–216.

Horgan, John, and Max Taylor, "Playing the 'Green Card': Financing the Provisional IRA: Part 1," *Terrorism and Political Violence*, Vol. 11, No. 2, 1999, pp. 1–38.

———, "Playing the 'Green Card': Financing the Provisional IRA: Part 2," *Terrorism and Political Violence*, Vol. 15, No. 2, 2003, pp. 1–60.

"Internet Makes Drug Traffickers Hard to Catch: DEA," Reuters, March 18, 2004.

Internet World Stats, "Internet Usage Statistics—The Big Picture," *Internet World Stats: Usage and Population Statistics*, undated Web page. As of January 15, 2007:
http://www.Internetworldstats.com/stats.htm

"Iraqi Improvised Explosive Device Attack," undated video hosted by Site Institute. As of February 20, 2007:
http://www.siteinstitute.org/multimedia/video/video_1093200298.wmv

Jackson, Brian, "Technology Acquisition by Terrorist Groups: Threat Assessment Informed by Lessons from Private Sector Technology Adoption," *Studies in Conflict and Terrorism*, Vol. 24, No. 3, May 1, 2001, pp. 183–214.

———, "Provisional Irish Republican Army," in Brian A. Jackson, John C. Baker, Peter Chalk, Kim Cragin, John V. Parachini, and Horacio R. Trujillo, *Aptitude for Destruction*, Vol. 2: *Case Studies of Organizational Learning in Five Terrorist Groups*, Santa Monica, Calif.: RAND Corporation, MG-332-NIJ, 2005, pp. 93–140. As of January 8, 2007:
http://www.rand.org/pubs/monographs/MG332/

———, *Technology Strategies for Homeland Security: Adaptation and Coevolution of Offense and Defense*, unpublished RAND research, 2006a.

———, "Training for Urban Resistance: The Case of the Provisional Irish Republican Army," in James J. F. Forest, ed., *The Making of a Terrorist: Recruitment, Training, and Root Causes*, Westport, Conn.: Praeger Security International, 2006b, pp. 119–135.

Jackson, Brian A., John C. Baker, Peter Chalk, Kim Cragin, John V. Parachini, and Horacio R. Trujillo, *Aptitude for Destruction*, Vol. 1: *Organizational Learning in Terrorist Groups and Its Implication for Combating Terrorism*, Santa Monica, Calif.: RAND Corporation, MG-331-NIJ, 2005a. As of January 8, 2007:
http://www.rand.org/pubs/monographs/MG331/

————, *Aptitude for Destruction,* Vol. 2: *Case Studies of Organizational Learning in Five Terrorist Groups,* Santa Monica, Calif.: RAND Corporation, MG-332-NIJ, 2005b. As of January 8, 2007:
http://www.rand.org/pubs/monographs/MG332/

Jackson, Brian A., Peter Chalk, Kim Cragin, Bruce Newsome, John V. Parachini, William Rosenau, Erin M. Simpson, Melanie Sisson, and Donald Temple, *Breaching the Fortress Wall: Understanding Terrorist Efforts to Overcome Defensive Technologies,* Santa Monica, Calif.: RAND Corporation, MG-481-DHS, 2007. As of July 23, 2007:
http://www.rand.org/pubs/monographs/MG481/

Jakobsson, Mikael, and T. L. Taylor, "The Sopranos Meets EverQuest: Social Networking in Massively Multiplayer Online Games," paper presented at melbourneDAC, the Fifth International Digital Arts and Culture Conference, Melbourne, Australia, May 19–23, 2003. As of January 8, 2007:
http://hypertext.rmit.edu.au/dac/papers/Jakobsson.pdf

Jarman, Neil, "Painting Landscapes: The Place of Murals in the Symbolic Construction of Urban Space," in Anthony D. Buckley, ed., *Symbols in Northern Ireland,* Belfast: Institute of Irish Studies, Queen's University of Belfast, 1998.

Jenkins, Brian Michael, *High Technology Terrorism and Surrogate War: The Impact of New Technology on Low-Level Violence,* Santa Monica, Calif.: RAND Corporation, P-5339, 1975. As of January 15, 2007:
http://www.rand.org/pubs/papers/P5339/

Kavkaz Center, "Video," undated Web page. As of January 15, 2007:
http://www.kavkazcenter.com/eng/video/

Kelley, Jack, "Terror Groups Hide Behind Web Encryption," *USA Today,* February 5, 2001. As of January 10, 2007:
http://www.usatoday.com/tech/news/2001-02-05-binladen.htm

Kennedy, John, "1Gbps Wireless Broadband Trials to Begin," *siliconrepublic.com,* August 9, 2005. As of January 15, 2007:
http://www.siliconrepublic.com/news/news.nv?storyid=single5205

Lamb, Christina, "Terrorist Video Shows Training for Hotel Attack," *The Daily Telegraph* (London), November 16, 2002. As of January 8, 2007:
http://www.telegraph.co.uk/news/main.jhtml?xml=/news/2002/11/17/wyemen317.xml

Lau, Stephen, "An Analysis of Terrorist Groups' Potential Use of Electronic Steganography," Bethesda, Md.: SANS Institute, February 18, 2003. As of January 10, 2007:
http://www.sans.org/reading_room/whitepapers/stenganography/554.php

Leader, Stefan H., and Peter Probst, "The Earth Liberation Front and Environmental Terrorism," *Terrorism and Political Violence,* Vol. 15, No. 4, 2003, pp. 37–58.

"Leaders—Mobile Phones and Development: Less Is More," *The Economist,* Vol. 376, No. 8434, 2005, p. 11.

Lewis, Nick, "Dangerous Games: How the Seductive Power of Video Games Is Being Harnessed to Push Deadly Agendas," *The Calgary Herald,* July 9, 2005, p. C11.

Loftus, Tom, "Virtual Worlds Wind Up in Real World's Courts," *MSNBC.com,* February 7, 2005. As of January 8, 2007:
http://www.msnbc.msn.com/id/6870901

Maxon, Terry, "Cell Phone Companies Add Tracking Abilities," *FSView and Florida Flambeau,* July 25, 2005. As of March 21, 2007:
http://media.www.fsunews.com/media/storage/paper920/news/2005/07/25/Lifestyles/Cell-Phone.Companies.Add.Tracking.Abilities-2355874.shtml

Merritt, Rick, "No Magic Bullet in Sight for UWB," *CommsDesign,* December 6, 2004. As of January 15, 2007:
http://www.commsdesign.com/news/market_nes/showArticle.jhtml?articleID=54800446

Mishra, Shitanshu, "Exploitation of Information and Communications Technology by Terrorist Organizations," *Strategic Analysis*, Vol. 27, No. 3, July–September 2003, pp. 439–462.

Mixed Reality Lab, "Human Pacman," August 5, 2006. As of January 15, 2007:
http://www.mixedreality.nus.edu.sg/index.php?option=com_content&task=view&id=42&Itemid=36

Mobile Pipeline, "Philips Platform Aims at Seamless Cell, Wi-Fi Handoffs," *CommWeb: The Telecommunications Pipeline*, March 14, 2005.

Molander, Roger C., B. David Mussington, and Peter A. Wilson, *Cyberpayments and Money Laundering: Problems and Promise*, Santa Monica, Calif.: RAND Corporation, MR-965-OSTP/FINCEN, 1998. As of January 8, 2007:
http://www.rand.org/pubs/monograph_reports/MR965/

Molander, Roger C., Andrew Riddile, and Peter A. Wilson, *Strategic Information Warfare: A New Face of War*, Santa Monica, Calif.: RAND Corporation, MR-661-OSD, 1996. As of January 8, 2007:
http://www.rand.org/pubs/monograph_reports/MR661/

"Money-Transfer Systems, Hawala Style," *CBC News Online*, June 11, 2004. As of January 8, 2007:
http://www.cbc.ca/news/background/banking/hawala.html

Morris, Chris, "The Greatest Story Never Played," *CNNMoney.com*, July 6, 2005. As of January 15, 2007:
http://money.cnn.com/2005/07/06/commentary/game_over/column_gaming/index.htm

Musgrove, Mike, "Intel Unveils Long-Range Wireless Technology," *The Washington Post*, April 19, 2005, p. E4. As of January 15, 2007:
http://www.washingtonpost.com/wp-dyn/articles/A63987-2005Apr18.html

National Commission on Terrorist Attacks upon the United States, *The 9/11 Commission Report: Final Report of the National Commission on Terrorist Attacks upon the United States*, New York: Norton, 2004. As of January 8, 2007:
http://www.gpoaccess.gov/911/index.html

National Communications System, *The Electronic Intrusion Threat to National Security and Emergency Preparedness (NS/EP) Internet Communications: An Awareness Document*, Arlington, Va.: Office of the Manager, National Communications System, 2000. As of January 15, 2007:
http://purl.access.gpo.gov/GPO/LPS18833

National Research Council, *Making the Nation Safer: The Role of Science and Technology in Countering Terrorism*, Washington, D.C.: National Academy Press, 2002.

Negroponte, Nicholas, *Being Digital*, New York: Knopf, 1995.

———, "Being Wireless," *Wired*, Vol. 10, No. 10, October 2002. As of January 15, 2007:
http://www.wired.com/wired/archive/10.10/wireless.html

Nelan, Bruce W., "How They Did It," *Time*, May 5, 1997, p. 149.

"Nokia Unveils RFID Phone Reader," *RFID Journal*, March 17, 2004. As of January 15, 2007:
http://www.rfidjournal.com/article/articleview/834/1/1/

Organisation for Economic Co-operation and Development, "OECD Broadband Statistics, December 2004." As of January 15, 2007:
http://www.oecd.org/document/60/0,2340,en_2649_201185_2496764_1_1_1_1,00.html

O'Callaghan, Sean, *The Informer*, London, UK: Corgi, 1999.

Parachini, John V., "Aum Shinrikyo," in Brian A. Jackson, John C. Baker, Peter Chalk, Kim Cragin, John V. Parachini, and Horacio R. Trujillo, *Aptitude for Destruction*, Vol. 2: *Case Studies of Learning in Five Terrorist Groups*, Santa Monica, Calif.: RAND Corporation, 2004, pp. 11–36. As of January 8, 2007:
http://www.rand.org/pubs/monographs/MG332/

Patrizio, Andy, "Did Game Play Role in Suicide?" *Wired News*, April 3, 2002. As of January 8, 2007:
http://www.wired.com/news/games/0,2101,51490,00.html

Petty, Richard E., and John T. Cacioppo, *Attitudes and Persuasion: Classic and Contemporary Approaches*, Boulder, Colo.: Westview Press, 1996.

Post, Jerrold, "From Car Bombs to Logic Bombs: The Growing Threat from Information Terrorism," *Terrorism and Political Violence*, Vol. 12, No. 2, Summer 2000, pp. 97–122.

"Preparing and Employing a Landmine," undated video hosted by the Site Institute. As of February 20, 2007:
http://www.siteinstitute.org/multimedia/video/video_1089342039.wmv

Project for Excellence in Journalism, "The State of the News Media 2006: An Annual Report on American Journalism," 2006. As of January 15, 2007:
http://www.stateofthenewsmedia.org/2006/index.asp

Ramakrishna, Kumar, *"Constructing" the Jemaah Islamiyah Terrorist: A Preliminary Inquiry*, Singapore: Institute of Defence and Strategic Studies, Nanyang Technological University, 2004.

Raman, B., "Terrorism in Thailand: Tech and Tactics Savvy," South Asia Analysis Group Paper No. 1321, April 5, 2005. As of January 15, 2007:
http://www.saag.org/papers14/paper1321.html

Ranstrop, Magnus, "Hizbollah's Command Leadership: Its Structure, Decision-Making and Relationship with Iranian Clergy and Institutions," *Terrorism and Political Violence*, Vol. 6, No. 3, 1994, pp. 303–339.

Rathmell, Andrew, "Cyber-Terrorism: The Shape of Future Conflict?" *Journal of the Royal United Service Institution*, October 1997, pp. 40–46.

Raymond, Louis, Pierre-Andre Julien, Jean-Bernard Carriere, and Richard Lachance, "Managing Technological Change in Manufacturing SMEs: A Multiple Case Analysis," *International Journal of Technology Management*, Vol. 11, No. 3/4, 1996, pp. 270–285.

"Ridge Wants Tech Firms to Enlist in Terrorism Fight," *USA Today*, April 24, 2002. As of January 10, 2007:
http://www.usatoday.com/tech/news/2002/04/24/homeland-security.htm

Ronfeldt, David, John Arquilla, Graham Fuller, and Melissa Fuller, *The Zapatista "Social Netwar" in Mexico*, Santa Monica, Calif.: RAND Corporation, MR-994-A, 1998. As of January 15, 2007:
http://www.rand.org/pubs/monograph_reports/MR994/

Roos, Gina, "Sensor Touts Industry-First G-Selectable Capability," *Automotive Design Line*, May 11, 2005. As of January 15, 2007:
http://www.automotivedesignline.com/showArticle.jhtml?printableArticle=true&articleId=163101454

Roth, John, Douglas Greenburg, and Serena B. Wille, *Monograph on Terrorist Financing: Staff Report to the Commission*, Washington, D.C.: National Commission on Terrorist Attacks upon the United States, 2004. As of January 8, 2007:
http://purl.access.gpo.gov/GPO/LPS53198

Sawyer, Ben, *Serious Games: Improving Public Policy Through Game-Based Learning and Simulation*, undated. As of January 8, 2007:
http://www.wilsoncenter.org/index.cfm?topic_id=1414&fuseaction=topics.documents&group_id=10264

Schaffer, Marvin B., "The Missile Threat to Civil Aviation," *Terrorism and Political Violence*, Vol. 10, No. 3, 1998, pp. 70–82.

Schmid, Alex Peter, and Janny de Graaf, *Violence as Communication: Insurgent Terrorism and the Western News Media*, London and Beverly Hills: Sage, 1982.

Schneier, Bruce, *Secrets and Lies: Digital Security in a Networked World*, New York: John Wiley, 2000.

Schwartau, Winn, *Information Warfare: Chaos on the Electronic Superhighway*, New York: Thunder's Mouth Press, 1994.

Singapore Ministry of Home Affairs, *The Jemaah Islamiyah Arrests and the Threat of Terrorism: White Paper*, Singapore: Ministry of Home Affairs, Republic of Singapore, 2003.

SITE Institute, "The Search for International Terrorist Entities," undated homepage. As of January 10, 2007: http://www.siteinstitute.org/

Squitieri, Tom, "Cyberspace Full of Terror Targets," *USA Today*, May 5, 2002. As of January 15, 2007: http://www.usatoday.com/tech/news/2002/05/06/cyber-terror.htm

Steinkuehler, Constance A., "Online Learning Studies," unpublished, undated research.

Stitt, Jason, and Les Chappell, "Games That Make Leaders: Top Researchers on the Rise of Play in Business and Education," *Wisconsin Technology Network*, January 20, 2005. As of January 8, 2007: http://wistechnology.com/article.php?id=1504

Sui, Hongfei, Jianxin Wang, Jianer Chen, and Songqiao Chen, "The Cost of Becoming Anonymous: On the Participant Payload in Crowds," *Information Processing Letters*, Vol. 90, No. 2, April 30, 2004, pp. 81–86.

Swamidass, P. M., and A. M. Waller, "A Classification of Approaches to Planning and Justifying New Manufacturing Technologies," *Journal of Manufacturing Systems*, Vol. 9, No. 3, 1990, pp. 181–193.

Taylor, Peter, "The New Al-Qaedo: Jihad.com," *BBC Television*, July 20, 2005. As of January 8, 2007: http://news.bbc.co.uk/2/hi/programmes/4683403.stm

TechWeb News, "NTT DoCoMo Ships Wi-FI/Cellular Phone," *Wireless Net Design Line*, November 17, 2004. As of January 15, 2007: http://www.wirelessnetdesignline.com/news/53700036

Thibodeau, Patrick, "FTC Examines Privacy Issues Raised by Data Collectors," *ComputerWorld*, March 26, 2001. As of January 8, 2007: http://www.computerworld.com/governmenttopics/government/policy/story/0,10801,58920,00.html

———, "Clarke: Terrorists Used Net for Info on Targets," *CNN.com*, February 15, 2002. As of January 8, 2007: http://archives.cnn.com/2002/TECH/internet/02/15/terrorists.internet.idg/

Thomas, Timothy L., "Al Qaeda and the Internet: The Danger of 'Cyberplanning,'" *Parameters*, Vol. 33, No. 3, Spring 2003, pp. 112–123.

United States v. Batiste, Abraham, Phanor, Herrera, Augustin, Lemorin, and Augustine, S.D. Fla., June 22, 2006. As of March 5, 2007: http://www.justice.gov/opa/documents/cts_batiste_indictment.pdf

United States v. Mokhtar Haouari, S4 00 Cr. 15 (JFK), S.D.N.Y., July 3, 2001.

United States v. Usama bin Laden, S(7)98 Cr. 1023, S.D.N.Y., February 21, 2001.

U.S. Army, *America's Army*, undated Web page. As of January 8, 2007: http://www.americasarmy.com/

U.S. Army Training and Doctrine Command, "A Military Guide to Terrorism in the Twenty-First Century," Fort Leavenworth, Kan.: U.S. Army Training and Doctrine Command, Deputy Chief of Staff for Intelligence, 2005. As of January 15, 2007: http://www.au.af.mil/au/awc/awcgate/army/guidterr/

U.S. Department of Commerce, National Oceanic and Atmospheric Administration, "NOAA ARL HYSPLIT Model," undated Web page. As of January 10, 2007: http://www.arl.noaa.gov/ready/hysplit4.html

U.S. Department of Homeland Security, *Information Bulletin, Terrorist Target Selection*, Washington, D.C., July 2005. Government publication; not releasable to the general public.

U.S. Department of Justice, *Al Qaeda Training Manual*, undated. As of January 10, 2007: http://www.au.af.mil/au/awc/awcgate/terrorism/alqaida%5Fmanual/

U.S. Environmental Protection Agency, "Computer-Aided Management of Emergency Operations (CAMEO®)," Web page, February 9, 2006. As of January 10, 2007:
http://www.epa.gov/ceppo/cameo/cameo.htm

U.S. General Accounting Office, *Information Security: Computer Attacks at Department of Defense Pose Increasing Risks: Report to Congressional Requesters*, Washington, D.C.: U.S. General Accounting Office, 1996.

———, *Information Security: Serious Weaknesses Place Critical Federal Operations and Assets at Risk: Report to the Committee on Governmental Affairs, U.S. Senate*, Washington, D.C.: U.S. General Accounting Office, 1998. As of January 15, 2007:
http://purl.access.gpo.gov/GPO/LPS17487

U.S. Secret Service, Carnegie Mellon University, and CERT Coordination Center, *2004 E-Crime Watch Survey: Summary of Findings*, Pittsburgh, Pa.: CERT, 2004. As of January 15, 2007:
http://www.cert.org/archive/pdf/2004eCrimeWatchSummary.pdf

Wechsler, William F., "Terror's Money Trail," *The New York Times*, September 26, 2001, p. A19.

Weimann, Gabriel, *Www.Terror.Net: How Modern Terrorism Uses the Internet*, Washington, D.C.: U.S. Institute of Peace, March 2004. As of January 8, 2007:
http://purl.access.gpo.gov/GPO/LPS47607

Wells, H. G., *The War of the Worlds*, London: William Heinemann, 1898.

Whine, Michael, "Cyberspace—A New Medium for Communication, Command, and Control by Extremists," *Studies in Conflict and Terrorism*, Vol. 22, No. 3, August 1, 1999, pp. 231–246.

Wilkinson, Paul, "The Media and Terrorism: A Reassessment," *Terrorism and Political Violence*, Vol. 9, No. 2, 1997, pp. 51–64.

Wilson, Peter A., and Roger C. Molander, *Exploring Money Laundering Vulnerabilities Through Emerging Cyberspace Technologies: A Caribbean-Based Exercise*, Santa Monica, Calif.: RAND Corporation, MR-1005-OSTP/FINCEN, 1998. As of January 8, 2007:
http://www.rand.org/pubs/monograph_reports/MR1005/

Windrem, Robert, "The Frightening Evolution of al-Qaida," *Dateline NBC*, June 24, 2005. As of January 10, 2007:
http://www.msnbc.msn.com/id/8307333/

Woodcock, Bruce Sterling, "An Analysis of MMOG Subscription Growth," *mmogchart.com*, undated Web page. As of January 15, 2007:
http://www.mmogchart.com

Yee, Nick, "The Demographics, Motivations, and Derived Experiences of Users of Massively Multi-User Online Graphical Environments," *Presence: Teleoperators and Virtual Environments*, Vol. 15, No. 3, 2006a, pp. 309–329.

———, "The Psychology of Massively Multi-User Online Role-Playing Games: Motivations, Emotional Investment, Relationships and Problematic Usage," in Ralph Schroeder and Ann-Sofie Axelsson, eds., *Avatars at Work and Play: Collaboration and Interaction in Shared Virtual Environments*, Dordrecht, The Netherlands: Springer, 2006b, pp. 187–208.

Zanini, Michele, "Middle Eastern Terrorism and Netwar," *Studies in Conflict and Terrorism*, Vol. 22, No. 3, August 1, 1999, pp. 247–256.

Zanini, Michele, and Sean J. A. Edwards, "The Networking of Terror in the Information Age," in John Arquilla and David Ronfeldt, eds., *Networks and Netwars: The Future of Terror, Crime, and Militancy*, Santa Monica, Calif.: RAND Corporation, MR-1382-OSD, 2001, pp. 29–60.

Zimbardo, Philip G., and Cynthia F. Hartley, "Cults Go to High School: A Theoretical and Empirical Analysis of the Initial Stage in the Recruitment Process," *Cultic Studies Journal*, Vol. 2, No. 1, 1985, pp. 91–147.

Zimbardo, Philip G., and Michael R. Leippe, *The Psychology of Attitude Change and Social Influence*, Philadelphia, Pa.: Temple University Press, 1991.